Digital Signal Processing with Python Programming

Digital Signal Processing with Python Programming

Maurice Charbit

WILEY

First published 2017 in Great Britain and the United States by ISTE Ltd and John Wiley & Sons, Inc.

Apart from any fair dealing for the purposes of research or private study, or criticism or review, as permitted under the Copyright, Designs and Patents Act 1988, this publication may only be reproduced, stored or transmitted, in any form or by any means, with the prior permission in writing of the publishers, or in the case of reprographic reproduction in accordance with the terms and licenses issued by the CLA. Enquiries concerning reproduction outside these terms should be sent to the publishers at the undermentioned address:

ISTE Ltd
27-37 St George's Road
London SW19 4EU
UK

www.iste.co.uk

John Wiley & Sons, Inc.
111 River Street
Hoboken, NJ 07030
USA

www.wiley.com

© ISTE Ltd 2017

The rights of Maurice Charbit to be identified as the author of this work have been asserted by him in accordance with the Copyright, Designs and Patents Act 1988.

Library of Congress Control Number: 2016955620

British Library Cataloguing-in-Publication Data
A CIP record for this book is available from the British Library
ISBN 978-1-78630-126-0

Contents

Preface ix

Notations and Abbreviations xi

A Few Functions of *Python*® xiii

Chapter 1. Useful Maths 1

    1.1. Basic concepts on probability 1
    1.2. Conditional expectation 10
    1.3. Projection theorem 11
        1.3.1. Conditional expectation 14
    1.4. Gaussianity 14
        1.4.1. Gaussian random variable 14
        1.4.2. Gaussian random vectors 15
        1.4.3. Gaussian conditional distribution 16
    1.5. Random variable transformation 18
        1.5.1. General expression 18
        1.5.2. Law of the sum of two random variables 19
        1.5.3. δ-method 20
    1.6. Fundamental theorems of statistics 22
    1.7. A few probability distributions 24

Chapter 2. Statistical Inferences 29

    2.1. First step: visualizing data 29
        2.1.1. Scatter plot 29
        2.1.2. Histogram/boxplot 30
        2.1.3. Q-Q plot 32
    2.2. Reduction of dataset dimensionality 34

    2.2.1. PCA 34
    2.2.2. LDA 36
  2.3. Some vocabulary 40
    2.3.1. Statistical inference 40
  2.4. Statistical model 41
    2.4.1. Notation 42
  2.5. Hypothesis testing 43
    2.5.1. Simple hypotheses 45
    2.5.2. Generalized likelihood ratio test (GLRT) 50
    2.5.3. χ^2 goodness-of-fit test 57
  2.6. Statistical estimation 58
    2.6.1. General principles 58
    2.6.2. Least squares method 62
    2.6.3. Least squares method for the linear model 64
    2.6.4. Method of moments 81
    2.6.5. Maximum likelihood approach 84
    2.6.6. Logistic regression 100
    2.6.7. Non-parametric estimation of probability distribution 103
    2.6.8. Bootstrap and others 107

Chapter 3. Inferences on HMM 113

  3.1. Hidden Markov models (HMM) 113
  3.2. Inferences on HMM 116
  3.3. Filtering: general case 117
  3.4. Gaussian linear case: Kalman algorithm 118
    3.4.1. Kalman filter 118
    3.4.2. RTS smoother 127
  3.5. Discrete finite Markov case 129
    3.5.1. Forward-backward formulas 130
    3.5.2. Smoothing formula at one instant 133
    3.5.3. Smoothing formula at two successive instants 134
    3.5.4. HMM learning using the EM algorithm 135
    3.5.5. The Viterbi algorithm 137

Chapter 4. Monte-Carlo Methods 141

  4.1. Fundamental theorems 141
  4.2. Stating the problem 141
  4.3. Generating random variables 144
    4.3.1. The cumulative function inversion method 144
    4.3.2. The variable transformation method 147
    4.3.3. Acceptance-rejection method 149
    4.3.4. Sequential methods 151
  4.4. Variance reduction 156

    4.4.1. Importance sampling 156
    4.4.2. Stratification 160
    4.4.3. Antithetic variates 164

Chapter 5. Hints and Solutions 167

   5.1. Useful maths 167
   5.2. Statistical inferences 170
   5.3. Inferences on HMM 226
   5.4. Monte-Carlo methods 251

Bibliography 261

Index 263

Preface

This book addresses the fundamental bases of statistical inferences. We shall presume throughout that readers have a good working knowledge of *Python®* language and of the basic elements of digital signal processing.

The most recent version is *Python®* 3.x, but many people are still working with *Python®* 2.x versions. All codes provided in this book work with both these versions. The official home page of the *Python®* Programming Language is https://www.python.org/. *Spyder®* is a useful open-source integrated dèvelopment environment (IDE) for programming in the *Python®* language. Briefly, we suggest to use the *Anaconda Python distribution*, which includes both *Python®* and *Spyder®*. The *Anaconda Python distribution* is located at https://www.continuum.io/downloads/.

The large part of the examples given in this book mainly use the modules numPy, which provides powerful numerical arrays objects, Scipy with high-level data processing routines, such as optimization, regression, interpolation and Matplotlib for plotting curves, histograms, Box and Whiskers plots, etc. See a list of useful functions p. xiii.

A brief outline of the contents of the book is given below.

Useful maths

In the first chapter, a short review of probability theory is presented, focusing on conditional probability, projection theorem and random variable transformation. A number of statistical elements will also be presented, including the great number law and the limit-central theorem.

Statistical inferences

The second chapter is devoted to statistical inference. Statistical inference consists of deducing some features of interest from a set of observations to a certain confidence level of reliability. This refers to a variety of techniques. In this chapter, we mainly focus on hypothesis testing, regression analysis, parameter estimation and determination of confidence intervals. Key notions include the Cramer–Rao bound, the Neyman–Pearson theorem, likelihood ratio tests, the least squares method for linear models, the method of moments and the maximum likelihood approach. The least squares method is a standard approach in regression analysis, and it is discussed in detail.

Inferences on HMM

In many problems, the variables of interest are only partially observed. Hidden Markov models (HMM) are well suited to accommodate this kind of problem. Their applications cover a wide range of fields, such as speech processing, handwriting recognition, the DNA analysis and monitoring and control. There are several issues with HMM inference. The key algorithms are the well-known Kalman filter, the Baum–Welch algorithm and the Viterbi algorithm to list only the most famous ones.

Monte-Carlo methods

Monte-Carlo methods refer to a broad class of algorithms that serve to perform quantities of interest. Typically, the quantities are integrals, i.e. the expectations of a given function. The key idea is using random sequences instead of deterministic sequences to achieve this result. The main issues are first the choice of the most appropriate random mechanism and, second, how to generate such a mechanism. In Chapter 4, the acceptance–rejection method, the Metropolis–Hastings algorithm, the Gibbs sampler, the importance sampling method, etc., are presented.

Maurice CHARBIT
October 2016

Notations and Abbreviations

$\emptyset$		empty set
$\mathbb{1}_A(x)$	=	$\begin{cases} 1 \text{ when } x \in A \\ 0 \text{ otherwise} \end{cases}$
(a,b]	=	$\{x : a < x \leq b\}$
$\delta(t)$		$\begin{cases} \text{Dirac distribution when } t \in \mathbb{R} \\ \text{Kronecker symbol when } t \in \mathbb{Z} \end{cases}$
Re(z)		real part of z
Im(z)		imaginary part of z
i or j	=	$\sqrt{-1}$
I_N		identity matrix of size N
A^*		complex conjugate of A
A^T		transpose of A
A^H		transpose-conjugate of A
A^{-1}		inverse matrix of A
$A^{\#}$		pseudo-inverse matrix of A
r.v./rv		random variable
$\mathbb{P}$		probability measure
$\mathbb{P}_\theta$		probability measure indexed by θ
$\mathbb{E}\{X\}$		expectation of X
$\mathbf{E}_\theta X$		expectation of X under $\mathbb{P}_\theta$
$X_c = X - \mathbb{E}\{X\}$		zero-mean random variable
$\text{var}(X) = \mathbb{E}\{|X_c|^2\}$		variance of X
$\text{cov}(X,Y) = \mathbb{E}\{X_c Y_c^*\}$		covariance of (X,Y)
$\text{cov}(X) = \text{cov}(X,X) = \text{var}(X)$		variance of X

$\mathbb{E}\{X\|Y\}$	conditional expectation of X given Y
$a \xrightarrow{d} b$	a converges in distribution to b
$a \xrightarrow{P} b$	a converges in probability to b
$a \xrightarrow{a.s.} b$	a converges almost surely to b
d.o.f.	degree of freedom
ARMA	AutoRegressive Moving Average
AUC	Area Under the ROC curve
c.d.f.	Cumulative Density Function
CRB	Cramer Rao Bound
EM	Expectation Maximization
GLRT	Generalized Likelihood Ratio Test
GEM	Generalized Expectation Maximization
GMM	Gaussian Mixture Model
HMM	Hidden Markov Model
i.i.d./iid	independent and identically distributed
LDA	Linear Discriminant Analysis
MC	Monte-Carlo
MLE	Maximum Likelihood Estimator
MME	Moment Method Estimator
MSE	Mean Square Error
OLS	Ordinary Least Squares
PCA	Principal Component Analysis
p.d.f.	Probability Density Function
ROC	Receiver Operational Characteristic
SNR	Signal to Noise Ratio
WLS	Weighted Least Squares

A Few Functions of *Python*®

To get function documentation, use `.__doc__`, e.g. `print(range.__doc__)`, or help, e.g. `help(zeros)` or `help('def')`, or `?`, e.g. `range.count?`.

- `def`: introduces a function definition

- `if, else, elif`: an if statement consists of a Boolean expression followed by one or more statements

- `for`: executes a sequence of statements multiple times

- `while`: repeats a statement or group of statements while a given condition is true

- `1j` or `complex`: returns complex value, e.g. `a=1.3+1j*0.2` or `a=complex(1.3,0.2)`

Methods:

- type `A=array([0,4,12,3])`, then type `A.` and tab, it follows a lot of methods, e.g. the argument of the maximum using `A.argmax`. For help type, e.g. `A.dot?`.

Functions:

- `int`: converts a number or string to an integer
- `len`: returns the number of items in a container
- `range`: returns an object that produces a sequence of integers
- `type`: returns the object type

From numpy:

- abs: returns the absolute value of the argument
- arange: returns evenly spaced values within a given interval
- argwhere: finds the indices of array elements that are non-zero, grouped by element
- array: creates an array
- cos, sin, tan: respectively calculate the cosine, the sine and the tangent
- cosh: calculates the hyperbolic cosine
- cumsum: calculates the cumulative sum of array elements
- diff: calculates the n-th discrete difference along a given axis
- dot: product of two arrays
- exp, log: respectively calculate the exponential, the logarithm
- fft: calculates the fft
- isinf: tests element-wise for positive or negative infinity
- isnan: tests element-wise for nan
- linspace: returns evenly spaced numbers over a specified interval
- loadtxt: loads data from a text file
- matrix: returns a matrix from an array-like object, or from a string of data
- max: returns the maximum of an array or maximum along an axis
- mean, std: respectively return the arithmetic mean and the standard deviation
- min: returns the minimum of an array or maximum along an axis
- nanmean, nanstd: respectively return the arithmetic mean and the standard deviation along a given axis while ignoring NaNs
- nansum: sum of array elements over a given axis, while ignoring NaNs
- ones: returns a new array of given shape and type, filled with ones
- pi: 3.141592653589793
- setdiff1d: returns the sorted, unique values of one array that are not in the other
- size: returns the number of elements along a given axis

- sort: returns a sorted copy of an array
- sqrt: computes the positive square-root of an array
- sum: sum of array elements over a given axis
- zeros: returns a new array of given shape and type, filled with zeroes

From numpy.linalg:
- eig: computes the eigenvalues and right eigenvectors of a square array
- pinv: computes the (Moore–Penrose) pseudo-inverse of a matrix
- inv: computes the (multiplicative) inverse of a matrix
- svd: computes Singular Value Decomposition

From numpy.random:
- rand: draws random samples from a uniform distribution over $(0, 1)$
- randn: draws random samples from the "standard normal" distribution
- randint: draws random integers from 'low' (inclusive) to 'high' (exclusive)

From scipy:

(for the random distributions, use the methods .pdf, .cdf, .isf, .ppf, etc.)
- norm: Gaussian random distribution
- gamma: gamma random distribution
- f: Fisher random distribution
- t: Student's random distribution
- chi2: chi-squared random distribution

From scipy.linalg:
- sqrtm: computes matrix square root

From matplotlib.pyplot:
- box, boxplot, clf, figure, hist, legend, plot, show, subplot
- title, txt, xlabel, xlim, xticks, ylabel, ylim, yticks

Datasets:

- statsmodels.api.datasets.co2, statsmodels.api.datasets.nile, statsmodels.api.datasets.star98, statsmodels.api.datasets.heart
- sklearn.datasets.load_boston, sklearn.datasets.load_diabetes
- scipy.misc.ascent

From sympy:

- Symbol, Matrix, diff, Inverse, trace, simplify

1

Useful Maths

1.1. Basic concepts on probability

Without describing in detail the formalism of the probability theory, we simply remind the reader of useful concepts. However, we advise the reader to consult some of the many books with authority on the subject [BIL 12].

In probability theory, we consider a *sample space* Ω, which is the set of all possible *outcomes* ω, and a collection $\mathcal{F}$ of its subsets with a structure of σ-algebra, the elements of which are called the *events*.

DEFINITION 1.1 (Random variable).– A real random variable X is a (measurable) application from the Ω to $\mathbb{R}$:

$$X : \omega \in \Omega \mapsto x \in \mathbb{R} \quad [1.1]$$

DEFINITION 1.2 (Discrete random variable).– A random variable X is said to be discrete if it takes its values in a subset of $\mathbb{R}$, at the most countable. If $\{a_0, \ldots, a_n, \ldots\}$, where $n \in \mathbb{N}$, denotes this set of values, the probability distribution of X is characterized by the sequence:

$$p_X(n) = \mathbb{P}\{X = a_n\} \quad [1.2]$$

representing the probability that X is equal to the element a_n. These values are such that $0 \leq p_X(n) \leq 1$ and $\sum_{n \geq 0} p_X(n) = 1$.

This leads us to the probability for the random variable X to belong to the interval $]a, b]$. It is given by:

$$\mathbb{P}\{X \in]a, b]\} = \sum_{n \geq 0} p_X(n) \mathbb{1}(a_n \in]a, b]) \quad [1.3]$$

The *cumulative distribution function (cdf)* of the random variable X is defined, for $x \in \mathbb{R}$, by:

$$F_X(x) = \mathbb{P}\{X \leq x\} = \sum_{\{n : a_n \leq x\}} p_X(n) \qquad [1.4]$$
$$= \sum_{n \geq 0} p_X(n) \mathbb{1}(a_n \in]-\infty, x])$$

It is a monotonic increasing function, with $F_X(-\infty) = 0$ and $F_X(+\infty) = 1$. Its graph is a staircase function, with jumps located at a_n with amplitude $p_X(n)$.

DEFINITION 1.3 (q-quantiles).– The k-th q-quantiles, associated with a given cumulative function $F(x)$, are written as:

$$c_k = \min\{x : F(x) \geq k/q\} \qquad [1.5]$$

where k goes from 1 to $q-1$. Therefore, the number of q-quantiles is $q-1$.

The q-quantiles are the limits of the partition of the probability range into q intervals of equal probability $1/q$. For example, the 2-quantile is the *median*.

More specifically, we have:

DEFINITION 1.4 (Median).– The median of the random variable X is the value M such that the cumulative function satisfies $F_X(M) = 1/2$.

The following program performs the q-quantiles of the Gaussian distribution[1]. Each area under the probability density equals $1/q$.

```
# -*- coding: utf-8 -*-
"""
Created on Fri Aug 12 09:11:27 2016
****** gaussianquantiles
@author: maurice
"""
from numpy import linspace, arange
from scipy.stats import norm
from matplotlib import pyplot as plt
x = linspace(-3,3,100); y = norm.pdf(x); plt.clf(); plt.plot(x,y)
q = 5; Qqi = arange(1,q)/q; quantiles = norm.ppf(Qqi)
plt.hold('on')
for iq in range(q-1):
```

---

[1] In *Python*® scipy.stats.norm.isf(1-a) = scipy.stats.norm.ppf(a).

```
print('%i-th of the %i-quantiles is %4.3e'%(iq+1,q,quantiles
    [iq]))
plt.plot([quantiles[iq],quantiles[iq]],[0.0,norm.pdf(quantiles
    [iq])],':')
plt.hold('off');plt.title('eachareaisequalto%4.2f'%(1.0/q));
plt.show();
```

DEFINITION 1.5 (Two discrete random variables).– Let $\{X, Y\}$ be two discrete random variables, with respective sets of values $\{a_0, \ldots, a_n, \ldots\}$ and $\{b_0, \ldots, b_k, \ldots\}$. The joint probability distribution is characterized by the sequence of positive values:

$$p_{XY}(n,k) = \mathbb{P}\{X = a_n, Y = b_k\} \quad [1.6]$$

with $0 \leq p_{XY}(n,k) \leq 1$ and $\sum_{n \geq 0} \sum_{k \geq 0} p_{XY}(n,k) = 1$.

This definition can easily be extended to the case of a finite number of random variables.

PROPERTY 1.1 (Marginal probability distribution).– Let $\{X, Y\}$ be two discrete random variables with their joint probability distribution $p_{XY}(n,k)$. The respective marginal probability distributions of X and Y are written as:

$$\begin{cases} \mathbb{P}\{X = a_n\} = \sum_{k=0}^{+\infty} p_{XY}(n,k) \\ \mathbb{P}\{Y = b_k\} = \sum_{n=0}^{+\infty} p_{XY}(n,k) \end{cases} \quad [1.7]$$

DEFINITION 1.6 (Continuous random variable).– A random variable is said to be continuous[2] if its values belong to $\mathbb{R}$ and if, for any real numbers a and b, the probability that X belongs to the interval $]a, b]$ is given by:

$$\mathbb{P}\{X \in]a,b]\} = \int_a^b p_X(x)dx = \int_{-\infty}^{\infty} p_X(x)\mathbb{1}(x \in]a,b])dx \quad [1.8]$$

where $p_X(x)$ is a function that must be positive or equal to zero such that $\int_{-\infty}^{+\infty} p_X(x)dx = 1$. $p_X(x)$ is called the probability density function (pdf) of X.

For any $x \in \mathbb{R}$, the *cumulative distribution function* (*cdf*) of the random variable X is defined by:

$$F_X(x) = \mathbb{P}\{X \leq x\} = \int_{-\infty}^{x} p_X(u)du \quad [1.9]$$

---

[2] The exact expression says that the probability distribution of X is *absolutely continuous with respect to the Lebesgue measure*.

It is a monotonic increasing function with $F_X(-\infty) = 0$ and $F_X(+\infty) = 1$. Notice that $p_X(x)$ also represents the derivative of $F_X(x)$ with respect to x.

DEFINITION 1.7 (Two continuous random variables).– Let $\{X, Y\}$ be two random variables with possible values in $\mathbb{R}^2$. Their probability distribution is said to be continuous if, for any domain Δ of $\mathbb{R}^2$, the probability that the pair (X, Y) belongs to Δ is given by:

$$\mathbb{P}\{(X, Y) \in \Delta\} = \iint_\Delta p_{XY}(x, y) dx dy \qquad [1.10]$$

where the function $p_{XY}(x, y) \geq 0$, and such that:

$$\iint_{\mathbb{R}^2} p_{XY}(x, y) dx dy = 1$$

$p_{XY}(x, y)$ is called the joint probability density function of the pair $\{X, Y\}$.

PROPERTY 1.2 (Marginal probability distributions).– Let $\{X, Y\}$ be two continuous random variables with the joint probability distribution $p_{XY}(x, y)$. The respective marginal probability density functions of X and Y can be written as:

$$\begin{cases} p_X(x) &= \int_{-\infty}^{+\infty} p_{XY}(x, y) dy \\ p_Y(y) &= \int_{-\infty}^{+\infty} p_{XY}(x, y) dx \end{cases} \qquad [1.11]$$

It is also possible to have a mixed situation, where one of the two variables is discrete and the other is continuous. This leads to the following:

DEFINITION 1.8 (Mixed random variables).– Let X be a discrete random variable with possible values $\{a_0, \ldots, a_n, \ldots\}$ and Y a continuous random variable with possible values in $\mathbb{R}$. For any value a_n, and for any real number pair (a, b), the probability:

$$\mathbb{P}\{X = a_n, Y \in \,]a, b]\} = \int_a^b p_{XY}(n, y) dy \qquad [1.12]$$

where the function $p_{XY}(n, y)$, with $n \in \{0, \ldots, k, \ldots\}$ and $y \in \mathbb{R}$, is ≥ 0 and verifies $\sum_{n \geq 0} \int_\mathbb{R} p_{XY}(n, y) dy = 1$.

DEFINITION 1.9 (Two independent random variables).– Two random variables X and Y are said to be independent if and only if their joint probability distribution is the product of the marginal probability distributions. This can be expressed as:

– for two discrete random variables: $p_{XY}(n,k) = p_X(n)\,p_Y(k)$
– for two continuous random variables: $p_{XY}(x,y) = p_X(x)\,p_Y(y)$
– for two mixed random variables: $p_{XY}(n,y) = p_X(n)\,p_Y(y)$

where the marginal probability distributions are obtained using formulae [1.7] and [1.11].

It is worth noting that, knowing $p_{XY}(x,y)$, we can tell whether or not X and Y are independent. To do this, we need to calculate the marginal probability distributions and check that $p_{XY}(x,y) = p_X(x)p_Y(y)$. If that is the case, then X and Y are independent.

The generalization to more than two random variables is given by the following definition.

DEFINITION 1.10 (Independent random variables).– The random variables $\{X_0, \ldots, X_{n-1}\}$ are jointly independent, if and only if their joint probability distribution is the product of their marginal probability distributions. This can be expressed as

$$p_{X_0 X_1 \ldots X_{n-1}}(x_0, x_1, \ldots, x_{n-1}) = p_{X_0}(x_0) p_{X_1}(x_1) \ldots p_{X_{n-1}}(x_{n-1}) \quad [1.13]$$

where the marginal probability distributions are obtained as integrals with respect to $(n-1)$ variables, calculated from $p_{X_0 X_1 \ldots X_{n-1}}(x_0, x_1, \ldots, x_{n-1})$.

For example, the marginal probability distribution of X_0 has the following expression:

$$p_{X_0}(x_0) = \underbrace{\int \ldots \int}_{\mathbb{R}^{n-1}} p_{X_0 X_1 \ldots X_{n-1}}(x_0, x_1, \ldots, x_{n-1}) dx_1 \ldots dx_{n-1}$$

In practice, the following result is a simple method for determining whether or not random variables are independent:

PROPERTY 1.3.– If $p_{X_0 X_1 \ldots X_{n-1}}(x_0, x_1, \ldots, x_{n-1})$ is a product of n positive functions of the type $f_0(x_0)$, $f_1(x_1)$, $\ldots$, $f_{n-1}(x_{n-1})$, then the variables are independent.

It should be noted that, if n random variables are independent of one another, it does not necessarily mean that they are jointly independent.

DEFINITION 1.11 (Mathematical expectation).– Let X be a random variable and $f(x)$ a function. The mathematical expectation of $f(X)$ is the deterministic value denoted by $\mathbb{E}\{f(X)\}$ and defined as follows:

– for a discrete r.v. by: $\mathbb{E}\{f(X)\} = \sum_{n \geq 0} f(a_n) p_X(n)$,

– for a continuous r.v. by: $\mathbb{E}\{f(X)\} = \int_{\mathbb{R}} f(x) p_X(x) dx$,

That can be extended to any number of random variables, e.g. for two random variables $\{X, Y\}$ and a function $f(x, y)$, the definition is:

– for 2 discrete r.v., by: $\mathbb{E}\{f(X,Y)\} = \sum_{n \geq 0} \sum_{k \geq 0} f(a_n, b_k) p_{XY}(n, k)$

– for 2 continuous r.v. by: $\mathbb{E}\{f(X,Y)\} = \int_{\mathbb{R}} \int_{\mathbb{R}} f(x, y) p_{XY}(x, y) dx dy$.

provided that all expressions exist.

From [1.3] and [1.8], the probability for X to belong to (a, b) may be seen as the expectation of the indicator function $\mathbb{1}(X \in (a, b))$.

PROPERTY 1.4.– If $\{X_0, X_1, \ldots, X_{n-1}\}$ are jointly independent, then for any integrable functions $f_0, f_1, \ldots, f_{n-1}$:

$$\mathbb{E}\left\{\prod_{k=0}^{n-1} f_k(X_k)\right\} = \prod_{k=0}^{n-1} \mathbb{E}\{f_k(X_k)\} \qquad [1.14]$$

DEFINITION 1.12 (Characteristic function).– The characteristic function of the probability distribution of the random variables $\{X_0, X_1, \ldots, X_{n-1}\}$ is the function of $(u_0, \ldots, u_{n-1}) \in \mathbb{R}^n$ defined by:

$$\phi_{X_0 \ldots X_{n-1}}(u_0, \ldots, u_{n-1}) = \mathbb{E}\left\{e^{ju_0 X_0 + \cdots + ju_{n-1} X_{n-1}}\right\} = \mathbb{E}\left\{\prod_{k=0}^{n-1} e^{ju_k X_k}\right\} \qquad [1.15]$$

As $|e^{juX}| = 1$, the characteristic function exists and is continuous even if the moments $\mathbb{E}\{X^k\}$ do not exist. For example, the Cauchy probability distribution, the probability density function of which is $p_X(x) = 1/\pi(1 + x^2)$, has no moment and has the characteristic function $e^{-|u|}$. Notice that $|\phi_{X_1 \ldots X_n}(u_1, \ldots, u_n)| \leq \phi_X(0, \ldots, 0) = 1$.

THEOREM 1.1 (Fundamental).– The random variables $\{X_0, X_1, \ldots, X_{n-1}\}$ are independent if and only if, for any point $(u_0, u_1, \ldots, u_{n-1})$ of $\mathbb{R}^n$:

$$\phi_{X_0 \ldots X_{n-1}}(u_0, \ldots, u_{n-1}) = \prod_{k=0}^{n-1} \phi_{X_k}(u_k)$$

Notice that the characteristic function $\phi_{X_k}(u_k)$ of the marginal probability distribution of X_k can be directly calculated using [1.15]. We have $\phi_{X_k}(u_k) = \mathbb{E}\left\{e^{ju_k X_k}\right\} = \phi_{X_0 \ldots X_{n-1}}(0, \ldots, 0, u_k, 0, \ldots, 0)$.

DEFINITION 1.13 (Mean, variance).– The mean of the random variable X is defined as the first-order moment, i.e. $\mathbb{E}\left\{X\right\}$. If the mean is equal to zero, the random variable is said to be centered. The variance of the random variable X is the quantity defined by:

$$\operatorname{var}(X) = \mathbb{E}\left\{(X - \mathbb{E}\left\{X\right\})^2\right\} = \mathbb{E}\left\{X^2\right\} - (\mathbb{E}\left\{X\right\})^2 \qquad [1.16]$$

The variance is always positive, and its square root is called the standard deviation.

As an exercise, we are going to show that, for any constants a and b:

$$\mathbb{E}\left\{aX + b\right\} = a\mathbb{E}\left\{X\right\} + b \qquad [1.17]$$

$$\operatorname{var}(aX + b) = a^2 \operatorname{var}(X) \qquad [1.18]$$

PROOF.– Expression [1.17] is a direct consequence of the integral's linearity. From $Y = aX + b$ and expression [1.17], we get $\operatorname{var}(Y) = \mathbb{E}\left\{(Y - \mathbb{E}\left\{Y\right\})^2\right\} = \mathbb{E}\left\{a^2(X - \mathbb{E}\left\{X\right\})^2\right\} = a^2 \operatorname{var}(X)$. ∎

A generalization of these two results to random vectors (their components are random variables) will be given by property [1.7].

DEFINITION 1.14 (Covariance, correlation).– Let $\{X, Y\}$ be two random variables. The covariance of X and Y is the quantity defined by:

$$\operatorname{cov}(X, Y) = \mathbb{E}\left\{(X - \mathbb{E}\left\{X\right\})(Y - \mathbb{E}\left\{Y\right\})\right\} \qquad [1.19]$$
$$= \mathbb{E}\left\{XY\right\} - \mathbb{E}\left\{X\right\}\mathbb{E}\left\{Y\right\}$$

The correlation coefficient is the quantity defined by:

$$\rho(X, Y) = \frac{\operatorname{cov}(X, Y)}{\sqrt{\operatorname{var}(X)}\sqrt{\operatorname{var}(Y)}} \qquad [1.20]$$

By applying the Schwartz inequality, we get $|\rho(X, Y)| \leq 1$.

X and Y are said to be uncorrelated if $\operatorname{cov}(X, Y) = 0$, i.e. if $\mathbb{E}\left\{XY\right\} = \mathbb{E}\left\{X\right\}\mathbb{E}\left\{Y\right\}$, therefore $\rho(X, Y) = 0$.

DEFINITION 1.15 (Mean vector and covariance matrix).– Let $\{X_0, X_1, \ldots, X_{n-1}\}$ be n random variables with the respective means $\mathbb{E}\{X_i\}$. The mean vector is the n dimension vector with the means $\mathbb{E}\{X_i\}$ as its components. The covariance matrix C is the $n \times n$ matrix with the entry $C_{ij} = \text{cov}(X_i, X_j)$ for $0 \leq i \leq n-1$ and $0 \leq j \leq n-1$.

Using the matrix notation $X = \begin{bmatrix} X_0 \ldots X_{n-1} \end{bmatrix}^T$, the mean vector can be expressed as:

$$\mathbb{E}\{X\} = \begin{bmatrix} \mathbb{E}\{X_0\} \ldots \mathbb{E}\{X_{n-1}\} \end{bmatrix}^T$$

the covariance matrix can be expressed as:

$$C = \mathbb{E}\{(X - \mathbb{E}\{X\})(X - \mathbb{E}\{X\})^T\}$$
$$= \mathbb{E}\{XX^T\} - \mathbb{E}\{X\}\mathbb{E}\{X^T\} \qquad [1.21]$$

and the correlation matrix can be expressed as:

$$R = DCD \qquad [1.22]$$

with

$$D = \begin{bmatrix} C_{00}^{-1/2} & 0 & \ldots & 0 \\ \vdots & \ddots & \ddots & \vdots \\ \vdots & \ddots & \ddots & 0 \\ 0 & \ldots & 0 & C_{n-1,n-1}^{-1/2} \end{bmatrix} \qquad [1.23]$$

R is obtained by dividing each element C_{ij} of C by $\sqrt{C_{ii}C_{jj}}$, provided that $C_{ii} \neq 0$. Therefore, $R_{ii} = 1$ and $|R_{ij}| \leq 1$.

Notice that the diagonal elements of a covariance matrix represent the respective variances of the n random variables. They are therefore positive.

If random variables are uncorrelated, their covariance matrix is diagonal and their correlation matrix is the identity matrix.

PROPERTY 1.5 (Positivity of the covariance matrix).– Any covariance matrix is positive, meaning that for any vector $a \in \mathbb{C}^n$, we have $a^H C a \geq 0$.

PROPERTY 1.6 (Bilinearity of the covariance).– Let $\{X_0, X_1, \ldots, X_{m-1}\}$ and $\{Y_0, \ldots, Y_{n-1}\}$ be random variables, and $v_0, \ldots, v_{m-1}, w_0, \ldots, w_{n-1}$ be arbitrary constants. Hence:

$$\operatorname{cov}\left(\sum_{i=0}^{m-1} v_i X_i, \sum_{j=0}^{n-1} w_j Y_j\right) = \sum_{i=0}^{m-1}\sum_{j=0}^{n-1} v_i w_j \operatorname{cov}(X_i, Y_j) \qquad [1.24]$$

PROOF.– Indeed, let V and W be the vectors of components v_i and w_j, respectively, and $A = V^T X$ and $B = W^T Y$. By definition, $\operatorname{cov}(A, B) = \mathbb{E}\{(A - \mathbb{E}\{A\})(B - \mathbb{E}\{B\})\}$. Replacing A and B with their respective expressions and using $\mathbb{E}\{A\} = V^T \mathbb{E}\{X\}$ and $\mathbb{E}\{B\} = W^T \mathbb{E}\{Y\}$, we obtain, successively:

$$\operatorname{cov}(A, B) = \mathbb{E}\{V^T(X - \mathbb{E}\{X\})(Y - \mathbb{E}\{Y\})^T W\} = \sum_{i=0}^{m-1}\sum_{j=0}^{n-1} v_i w_j \operatorname{cov}(X_i, Y_j)$$

thus demonstrating expression [1.24]. ∎

Using matrix notation, expression [1.24] is written in the following form:

$$\operatorname{cov}\left(V^T X, W^T Y\right) = V^T C W \qquad [1.25]$$

where C designates the covariance matrix of X and Y.

PROPERTY 1.7 (Linear transformation of a random vector).– Let $\{X_0, \ldots, X_{n-1}\}$ be n random variables. We let X the random vector whose components are X_i, $\mathbb{E}\{X\}$ its mean vector and C_X its covariance matrix, and let $\{Y_0, \ldots, Y_{q-1}\}$ be q random variables obtained by the linear transformation:

$$\begin{bmatrix} Y_0 \\ \vdots \\ Y_{q-1} \end{bmatrix} = A \begin{bmatrix} X_0 \\ \vdots \\ X_{n-1} \end{bmatrix} + b$$

where A is a $q \times n$ matrix and b is a non-random vector with the adequate sizes. We then have:

$$\mathbb{E}\{Y\} = A\mathbb{E}\{X\} + b$$
$$C_Y = A C_X A^T$$

DEFINITION 1.16 (White sequence).– Let $\{X_0, \ldots, X_{n-1}\}$ be a set of n random variables. They are said to form a white sequence if $\operatorname{var}(X_i) = \sigma^2$ and if

$\text{cov}(X_i, X_j) = 0$ for $i \neq j$. Hence, their covariance matrix can be expressed as follows:

$$C = \sigma^2 I_n$$

where I_n is the $n \times n$ identity matrix.

PROPERTY 1.8 (Independence $\Rightarrow$ non-correlation).– Let $\{X_0, \ldots, X_{n-1}\}$ be n independent random variables, then they are uncorrelated. Usually, the converse statement is false.

1.2. Conditional expectation

DEFINITION 1.17 (Conditional expectation).– Let X be a random variable and Y a random vector taking values, respectively, in $\mathcal{X} \subset \mathbb{R}$ and $\mathcal{Y} \subset \mathbb{R}^q$. Let $p_{XY}(x,y)$ be their joint probability density. The conditional expectation of X, given Y, is a (measurable) real valued function $g(Y)$, such that, for any other real valued function $h(Y)$, we have:

$$\mathbb{E}\left\{|X - g(Y)|^2\right\} \leq \mathbb{E}\left\{|X - h(Y)|^2\right\} \qquad [1.26]$$

$g(Y)$ is commonly denoted by $\mathbb{E}\{X|Y\}$.

PROPERTY 1.9 (Conditional probability distribution).– We consider a random variable X and a random vector Y taking values, respectively, in $\mathcal{X} \subset \mathbb{R}$ and $\mathcal{Y} \subset \mathbb{R}^q$ with joint probability density $p_{XY}(x,y)$. Then, $\mathbb{E}\{X|Y\} = g(Y)$ with:

$$g(y) = \int_{\mathcal{X}} x \, p_{X|Y}(x,y) dx$$

where

$$p_{X|Y}(x,y) = \frac{p_{XY}(x,y)}{p_Y(y)} \text{ and } p_Y(y) = \int_{\mathcal{X}} p_{XY}(x,y) dx \qquad [1.27]$$

$p_{X|Y}(x,y)$ is called the conditional probability distribution of X given Y.

PROPERTY 1.10.– The conditional expectation verifies the following properties:

1) linearity: $\mathbb{E}\{a_1 X_1 + a_2 X_2 | Y\} = a_1 \mathbb{E}\{X_1|Y\} + a_2 \mathbb{E}\{X_2|Y\}$;

2) orthogonality: $\mathbb{E}\{(X - \mathbb{E}\{X|Y\})h(Y)\} = 0$ for any function $h : \mathcal{Y} \mapsto \mathbb{R}$;

3) $\mathbb{E}\{h(Y)f(X)|Y\} = h(Y)\mathbb{E}\{f(X)|Y\}$, for all functions $f : \mathcal{X} \mapsto \mathbb{R}$ and $h : \mathcal{Y} \mapsto \mathbb{R}$;

4) $\mathbb{E}\{\mathbb{E}\{f(X,Y)|Y\}\} = \mathbb{E}\{f(X,Y)\}$ for any function $f : \mathcal{X} \times \mathcal{Y} \mapsto \mathbb{R}$; specifically

$$\mathbb{E}\{\mathbb{E}\{X|Y\}\} = \mathbb{E}\{X\}$$

5) refinement by conditioning: it can be shown (see page 14) that

$$\text{cov}\left(\mathbb{E}\{X|Y\}\right) \leq \text{cov}\left(X\right) \qquad [1.28]$$

That has a clear meaning: the variance is reduced by conditioning;

6) if X and Y are independent, then $\mathbb{E}\{f(X)|Y\} = \mathbb{E}\{f(X)\}$. Specifically, $\mathbb{E}\{X|Y\} = \mathbb{E}\{X\}$. The reciprocal is not true;

7) $\mathbb{E}\{X|Y\} = X$, if and only if X is a function of Y.

1.3. Projection theorem

DEFINITION 1.18 (Dot product).– Let $\mathcal{H}$ be a vector space constructed over $\mathbb{C}$. The dot product is an application

$$X, Y \in \mathcal{H} \times \mathcal{H} \mapsto (X, Y) \in \mathbb{C}$$

which verifies the following properties:

– $(X, Y) = (Y, X)^*$;
– $(\alpha X + \beta Y, Z) = \alpha(X, Z) + \beta(Y, Z)$;
– $(X, X) \geq 0$. The equality occurs if and only if $X = 0$.

A vector space is a Hilbert space if it is complete with respect to its dot product[3]. The norm of X is defined by $\|X\| = \sqrt{(X, X)}$ and the distance between two elements is defined by $d(X, Y) = \|X - Y\|$. Two elements X and Y are said to be orthogonal, denoted $X \perp Y$, if and only if $(X, Y) = 0$. The demonstration of the following properties is trivial:

– Schwarz inequality:

$$|(X, Y)| \leq \|X\| \|Y\| \qquad [1.29]$$

the equality occurs if and only if there exists λ such that $X = \lambda Y$;

---

[3] A definition of the term "complete" in this context may be found in mathematical textbooks. In the context of our presentation, this property plays a concealed role, e.g. in the existence of the orthogonal projection in theorem 1.2.

– triangular inequality:

$$|\,\|X\| - \|Y\|\,| \leq \|X - Y\| \leq \|X\| + \|Y\| \qquad [1.30]$$

– parallelogram identity:

$$\|X + Y\|^2 + \|X - Y\|^2 = 2\|X\|^2 + 2\|Y\|^2 \qquad [1.31]$$

In a Hilbert space, the projection theorem enables us to associate any given element from the space with its best quadratic approximation contained in a closed vector subspace:

THEOREM 1.2 (Projection theorem).– Let $\mathcal{H}$ be a Hilbert space defined over $\mathbb{C}$ and $\mathcal{C}$ a closed sub-space of $\mathcal{H}$. Any vector X of $\mathcal{H}$ may then be associated with a unique element X_0 of $\mathcal{C}$, such that $\forall Y \in \mathcal{C}$ we have $d(X, X_0) \leq d(X, Y)$. Vector X_0 verifies, for any $Y \in \mathcal{C}$, the relationship $(X - X_0) \perp Y$.

The relationship $(X - X_0) \perp Y$ constitutes the *orthogonality principle*.

A geometric representation of the orthogonality principle is shown in Figure 1.1. The element of $\mathcal{C}$ closest in distance to X is given by the *orthogonal projection* of X onto $\mathcal{C}$. It follows that

$$\begin{aligned}\|X - X_0\|^2 &= (X, X - X_0) - (X_0, X - X_0) \\ &= \|X\|^2 - (X, X_0)\end{aligned} \qquad [1.32]$$

where the term $(X_0, X - X_0) = 0$ due to the orthogonality principle.

In what follows, the vector X_0 will be noted as $(X|\mathcal{C})$, or $(X|Y_{0:n-1})$ when the sub-space onto which projection occurs is spanned by the linear combinations of vectors $Y_0, \ldots, Y_{n-1}$.

The most simple application of theorem 1.2 provides that, for any vector $X \in \mathcal{H}$ and any vector $\varepsilon \in \mathcal{C}$:

$$(X|\varepsilon) = \frac{(X, \varepsilon)}{(\varepsilon, \varepsilon)} \varepsilon \qquad [1.33]$$

The projection theorem leads us to define an application associating element X with element $(X|\mathcal{C})$. This application is known as the *orthogonal projection* of X onto $\mathcal{C}$. The orthogonal projection verifies the following properties:

1) linearity: $(\lambda X_1 + \mu X_1|\mathcal{C}) = \lambda(X_1|\mathcal{C}) + \mu(X_2|\mathcal{C})$;
2) contraction: $\|(X|\mathcal{C})\| \leq \|X\|$;
3) if $\mathcal{C}' \subset \mathcal{C}$, then $((X|\mathcal{C})|\mathcal{C}') = (X|\mathcal{C}')$;
4) if $\mathcal{C}_1 \perp \mathcal{C}_2$, then $(X|\mathcal{C}_1 \oplus \mathcal{C}_2) = (X|\mathcal{C}_1) + (X|\mathcal{C}_2)$.

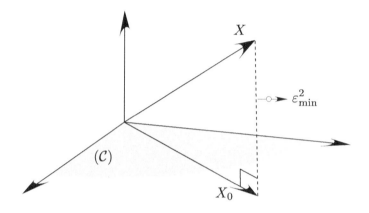

Figure 1.1. *Orthogonality principle: the point X_0 which is the closest to X in $\mathcal{C}$ is such that $X - X_0$ is orthogonal to $\mathcal{C}$*

The following result is fundamental:

$$(X|Y_{0:n}) = (X|Y_{0:n-1}) + (X|\varepsilon) = (X|Y_{0:n-1}) + \frac{(X,\varepsilon)}{(\varepsilon,\varepsilon)}\varepsilon \qquad [1.34]$$

where $\varepsilon = Y_n - (Y_n|Y_{0:n-1})$.

PROOF.– Indeed, as the sub-space spanned by $Y_{0:n}$ coincides with the sub-space spanned by $(Y_{0:,n-1}, \varepsilon)$ and as ε is orthogonal to the sub-space generated by $(Y_{0:n-1})$, then property [4] applies. To complete the proof, we use [1.33]. ∎

Formula [1.34] is the basic formula used in the determination of many recursive algorithms, such as Kalman filter or Levinson recursion.

THEOREM 1.3 (Square-integrable r.v.).– Let $\mathcal{L}_P^2$ be the vector space of square-integrable random variables, defined on the probability space $(\Omega, \mathcal{A}, P)$. Using the scalar product $(X, Y) = \mathbb{E}\{XY\}$, $\mathcal{L}_P^2$ has a Hilbert space structure.

1.3.1. *Conditional expectation*

The conditional expectation $\mathbb{E}\{X|Y\}$ may be seen as the orthogonal projection of X onto sub-space $\mathcal{C}$ of all measurable functions of Y. Similarly, $\mathbb{E}\{X\}$ may be seen as the orthogonal projection of X onto the sub-space $\mathcal{D}$ of the constant random variables. These vectors are shown in Figure 1.2. As $\mathcal{D} \subset \mathcal{C}$, using Pythagoras's theorem, we deduce that:

$$\text{var}(X) = \|X - \mathbb{E}\{X\}\|^2 = \|X - \mathbb{E}\{X|Y\}\|^2 + \underbrace{\|\mathbb{E}\{X|Y\} - \mathbb{E}\{X\}\|^2}_{=\text{var}(\mathbb{E}\{X|Y\})}$$

demonstrating $\text{var}(\mathbb{E}\{X|Y\}) \leq \text{var}(X)$. That can be extended to random vectors, giving the inequality [1.28], i.e. $\text{cov}(\mathbb{E}\{X|Y\}) \leq \text{cov}(X)$.

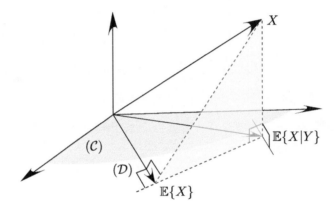

Figure 1.2. *The conditional expectation* $\mathbb{E}\{X|Y\}$ *is the orthogonal projection of X onto the set $\mathcal{C}$ of measurable functions of Y. The expectation $\mathbb{E}\{X\}$ is the orthogonal projection of X onto the set $\mathcal{D}$ of constant functions. Clearly, $\mathcal{D} \subset \mathcal{C}$*

1.4. Gaussianity

1.4.1. *Gaussian random variable*

DEFINITION 1.19.– A random variable X is said to be Gaussian, or normal, if all its values belong to $\mathbb{R}$ and if its characteristic function (see expression [1.15]) has the expression:

$$u \in \mathbb{R} \mapsto \phi_X(u) = \exp\left(jmu - \frac{1}{2}\sigma^2 u^2\right) \qquad [1.35]$$

where m is a real value and σ is a positive value. We check that its mean is equal to m and its variance to σ^2.

If $\sigma \neq 0$, it can be shown that the probability distribution has a probability density function with the expression:

$$p_X(x) = \frac{1}{\sigma\sqrt{2\pi}} \exp\left(-\frac{(x-m)^2}{2\sigma^2}\right) \qquad [1.36]$$

1.4.2. *Gaussian random vectors*

DEFINITION 1.20 (Gaussian vector).– $\{X_0, \ldots, X_{n-1}\}$ are said to be n jointly Gaussian variables, or that the length n vector $\begin{bmatrix} X_0 & \ldots & X_{n-1} \end{bmatrix}^T$ is Gaussian, if and only if any linear combination of its components, that is to say $Y = a^H X$ for any $a = \begin{bmatrix} a_0 & \ldots & a_{n-1} \end{bmatrix}^T \in \mathbb{C}^n$, is a Gaussian random variable.

THEOREM 1.4 (Distribution of a Gaussian vector).– It can be shown that the probability distribution of a Gaussian vector, with length n, mean vector m and an $(n \times n)$ covariance matrix C, has the characteristic function:

$$\phi_X(u) = \exp\left(jm^T u - \frac{1}{2} u^T C u\right) \qquad [1.37]$$

where $u = \begin{bmatrix} u_0 & \ldots & u_{n-1} \end{bmatrix}^T \in \mathbb{R}^n$. Let $x = \begin{bmatrix} x_0 & \ldots & x_{n-1} \end{bmatrix}^T$. If $\det\{C\} \neq 0$, the probability distribution's density has the expression:

$$p_X(x) = \frac{1}{(2\pi)^{n/2}\sqrt{\det\{C\}}} \exp\left(-\frac{1}{2}(x-m)^T C^{-1}(x-m)\right) \qquad [1.38]$$

In the following text, the Gaussian distributions will be denoted by $\mathcal{N}(m,C)$.

THEOREM 1.5.– (Gaussian case: non-correlation $\Rightarrow$ independence) If n jointly Gaussian variables are uncorrelated, C is diagonal, then they are independent.

THEOREM 1.6 (Linear transformation of a Gaussian vector).– Let $\begin{bmatrix} X_0 & \ldots & X_{n-1} \end{bmatrix}^T$ be a Gaussian vector with a mean vector m_X and a covariance matrix C_X. The random vector $Y = AX + b$, where A and b are a matrix and a vector, respectively, with the ad hoc length, is Gaussian and we have:

$$m_Y = Am_X + b \quad \text{and} \quad C_Y = AC_X A^T \qquad [1.39]$$

In other words, the Gaussian nature of a vector is untouched by linear transformations.

Equations [1.39] are a direct consequence of definition [1.20] and of property [1.7].

More specifically, if X is a random Gaussian vector $\mathcal{N}(m,C)$, then the random variable $Z = C^{-1/2}(X - M)$ follows a Gaussian distribution $\mathcal{N}(0, I)$. Another way of expressing this is to say that if Z has the distribution $\mathcal{N}(0, I)$, then $X = M + C^{1/2}Z$ has the distribution $\mathcal{N}(m,C)$.

Note that, if C denotes a positive matrix, a square root of C is a matrix M, which verifies:

$$C = MM^H \qquad [1.40]$$

Hence, if M is a square root of C, then for any unitary matrix U, i.e. such that $UU^H = I$, matrix MU is also a square root of C. Matrix M is therefore defined to be within a unitary matrix. One of the square roots is positive, and is obtained in *Python*® using the function numpy.sqrtm.

The Gaussian distribution is defined using the first- and second-order moments, i.e. the mean and the covariance. Consequently, all moments of an order greater than 2 are expressed as a function of the first two values. The following theorem covers the specific case of a moment of order 4.

THEOREM 1.7 (Moment of order 4).– Let X_1, X_2, X_3 and X_4 be four real *centered* Gaussian random variables. Hence,

$$\mathbb{E}\{X_1 X_2 X_3 X_4\} = \mathbb{E}\{X_1 X_2\}\mathbb{E}\{X_3 X_4\} \qquad [1.41]$$
$$+ \mathbb{E}\{X_1 X_3\}\mathbb{E}\{X_2 X_4\} + \mathbb{E}\{X_1 X_4\}\mathbb{E}\{X_2 X_3\}$$

Hence:

$$\begin{aligned} &\operatorname{cov}(X_1 X_2, X_3 X_4) \\ &= \mathbb{E}\{X_1 X_2 X_3 X_4\} - \mathbb{E}\{X_1 X_2\}\mathbb{E}\{X_3 X_4\} \\ &= \mathbb{E}\{X_1 X_3\}\mathbb{E}\{X_2 X_4\} + \mathbb{E}\{X_1 X_4\}\mathbb{E}\{X_2 X_3\} \end{aligned} \qquad [1.42]$$

1.4.3. *Gaussian conditional distribution*

Consider two jointly Gaussian random vectors X and Y, taking their values in $\mathbb{R}^p$ and $\mathbb{R}^q$, respectively. The respective means are noted $\mu_X \in \mathbb{R}^p$ and $\mu_Y \in \mathbb{R}^q$, and

$$C = \begin{bmatrix} \operatorname{cov}(X,X) & \operatorname{cov}(X,Y) \\ \operatorname{cov}(Y,X) & \operatorname{cov}(Y,Y) \end{bmatrix} \qquad [1.43]$$

is the joint covariance matrix, where cov (X, X) is a $p \times p$ positive matrix, cov (X, Y) is a $p \times q$ matrix and cov (Y, Y) is a $q \times q$ positive matrix. This leads to the following results:

PROPERTY 1.11.– The conditional expectation of X given Y coincides with the orthogonal projection of X onto the affine sub-space spanned by $\mathbf{1}$ and Y (this is written $B + AY$). Hence:

– the conditional expectation is expressed as:

$$\mathbb{E}\{X|Y\} = \mu_X + \text{cov}(X, Y)\left[\text{cov}(Y, Y)\right]^{-1}(Y - \mu_Y) \qquad [1.44]$$

– the conditional covariance is expressed as:

$$\text{cov}(X|Y) = \text{cov}(X, X) - \text{cov}(X, Y)\left[\text{cov}(Y, Y)\right]^{-1}\text{cov}(Y, X) \qquad [1.45]$$

– the conditional distribution of $p_{X|Y}(x, y)$ is Gaussian. Its mean is expressed as [1.44] and its covariance is given by expression [1.45].

PROOF.– Let $g(Y)$ be the second member of [1.44], and let us demonstrate that $g(Y)$ is the conditional expectation of X, given Y. A straightforward calculation shows that $\mathbb{E}\{(X - g(Y))Y^T\} = 0$. Consequently, the random vectors $Z = (X - g(Y))$ and Y are uncorrelated. As the vectors are jointly Gaussian, following property [1.10], they are independent and hence $\mathbb{E}\{Z|Y\} = \mathbb{E}\{Z\}$. Using the second member of [1.44], we obtain $\mathbb{E}\{Z\} = 0$. On the other hand:

$$\mathbb{E}\{X - g(Y)|Y\} = \mathbb{E}\{X|Y\} - g(Y)$$

It follows that $\mathbb{E}\{X|Y\} = g(Y)$. To demonstrate expression [1.45], take $X^c = X - \mu_X$ and $X_Y^c = \mathbb{E}\{X|Y\} - \mu_X$. Hence, successively:

$$\mathbb{E}\{(X^c - X_Y^c)((X^c - X_Y^c)^T|Y\} = \mathbb{E}\{(X^c - X_Y^c)(X^c - X_Y^c)^T\}$$
$$= \mathbb{E}\{(X^c - X_Y^c)X^{c,T}\}$$
$$= \text{cov}(X, X) - \text{cov}(X, Y)\left[\text{cov}(YY)\right]^{-1}\text{cov}(Y, X)$$

where, in the first equality, we use the fact that $(X^c - X_Y^c)$ is independent of Y. In conclusion, the conditional distribution of X, given Y, is written as:

$$p_{X|Y}(x, y) = \mathcal{N}\left(\mu_X + \text{cov}(X, Y)\left[\text{cov}(YY)\right]^{-1}(Y - \mu_Y),\right.$$
$$\left.\text{cov}(X, X) - \text{cov}(X, Y)\left[\text{cov}(YY)\right]^{-1}\text{cov}(Y, X)\right) \qquad [1.46]$$

Note that the distribution for random vector $\mathbb{E}\{X|Y\}$ should not be confused with the conditional distribution $p_{X|Y}(x, y)$ of X, given Y. We shall restrict ourselves to the scalar case, taking μ_X and μ_Y as the respective means of X and Y, and

$$C = \begin{bmatrix} \sigma_X^2 & \rho\sigma_X\sigma_Y \\ \rho\sigma_X\sigma_Y & \sigma_Y^2 \end{bmatrix}$$

with $-1 \leq \rho \leq 1$ as the covariance matrix. The conditional distribution of X given Y has a probability density $p_{X|Y}(x;y) = \mathcal{N}(\mu_X + \rho\sigma_X(y - \mu_Y)/\sigma_Y, \sigma_X^2(1 - \rho^2))$. On the other hand, the random variable distribution $\mathbb{E}\{X|Y\}$ has a probability density of $\mathcal{N}(\mu_X, \rho^2\sigma_X^2)$. Indeed, based on equation [1.44], $\mathbb{E}\{\mathbb{E}\{X|Y\}\} = \mu_X$ and $\mathbb{E}\{(\mathbb{E}\{X|Y\} - \mu_X)^2\} = \rho^2\sigma_X^2\sigma_Y^2/\sigma_Y^2 = \rho^2\sigma_X^2$.

1.5. Random variable transformation

1.5.1. *General expression*

In many cases, it is necessary to determine the distribution of $Y = g(X)$ from the distribution of X. In this section, we shall consider this question in the context of continuous random vectors with finite dimension.

Let us consider the one-to-one mapping[4]:

$$y = g(y) \in \mathbb{R}^d \Leftrightarrow x = h(y) \in \mathbb{R}^d$$

and a random vector X with probability density $p_X(x)$. In this case, the probability density of the random vector Y is given by:

$$p_Y(y) = p_X(h(y)) \left| \det\left\{ \frac{\partial x}{\partial y} \right\} \right| \qquad [1.47]$$

where $\frac{\partial x}{\partial y}$ denotes the *Jacobian* of h defined by:

$$\frac{\partial x}{\partial y} = \begin{bmatrix} \frac{\partial h_0(y_0, \ldots, y_{d-1})}{\partial y_0} & \cdots & \frac{\partial h_{d-1}(y_0, \ldots, y_{m-1})}{\partial y_0} \\ \vdots & & \\ \frac{\partial h_0(y_0, \ldots, y_{d-1})}{\partial y_{d-1}} & \cdots & \frac{\partial h_{d-1}(y_0, \ldots, y_{d-1})}{\partial y_{d-1}} \end{bmatrix}$$

---

[4] In cases where the transformation is not bijective, it is necessary to sum all of the solutions x, such that $y = g(x)$.

Note that the Jacobian of a bijective function has one particularly useful property. Taking a bijective function $x \in \mathbb{R}^d \leftrightarrow y \in \mathbb{R}^d$, we have:

$$\frac{\partial x}{\partial y} \times \frac{\partial y}{\partial x} = I_d \qquad [1.48]$$

The expression [1.48] allows us to calculate the Jacobian using the expression which is easiest to calculate, and, if necessary, to take the inverse.

1.5.2. *Law of the sum of two random variables*

As an example, let us consider two random variables X_0 and X_1 with a joint probability density $p_{X_0 X_1}(x_0, x_1)$. We wish to determine the joint distribution of the pair $\{Y_0, Y_1\}$, defined by the following transformation:

$$\begin{cases} Y_0 = X_0 \\ Y_1 = X_0 + X_1 \end{cases} \Leftrightarrow \begin{cases} X_0 = Y_0 \\ X_1 = Y_1 - Y_0 \end{cases}$$

We can verify that the determinant of the Jacobian is equal to 1. Applying [1.47], we obtain the following probability density for the pair $\{Y_0, Y_1\}$:

$$p_{Y_0 Y_1}(y_0, y_1) = p_{X_0 X_1}(y_0, y_1 - y_0)$$

From this, the probability density of $Y_1 = X_0 + X_1$ may be derived as the marginal distribution of Y_1. We obtain:

$$p_{Y_1}(y_1) = \int_{\mathbb{R}} p_{X_0 X_1}(y_0, y_1 - y_0) dy_0$$

In cases where X_0 and X_1 are independent:

$$p_{X_0 X_1}(x_0, x_1) = p_{X_0}(x_0) p_{X_1}(x_1)$$

hence:

$$p_{Y_1}(y_1) = \int_{\mathbb{R}} p_{X_0}(y_0) p_{X_1}(y_1 - y_0) dy_0$$

which is the expression of the convolution product $(p_{X_0} \star p_{X_1})(y_1)$.

EXERCISE 1.1 (Module and phase joint law of a 2D Gaussian r.v. (see p. 61)).– Consider two independent, centered Gaussian random variables (X, Y) of the same variance σ^2. We are concerned with the bijective variable transformation $(X, Y) \mapsto (R, \Theta)$ defined by

$$\begin{cases} r = \sqrt{x^2 + y^2} & r \in \mathbb{R}^+ \\ \theta = \arg(x + jy) & \theta \in (0, 2\pi) \end{cases} \Leftrightarrow \begin{cases} x = r\cos(\theta) & X \in \mathbb{R} \\ y = r\sin(\theta) & Y \in \mathbb{R} \end{cases}$$

Determine the joint distribution of (R, Θ) and the marginal distributions of R and Θ.

1.5.3. δ-method

In cases where no closed-form expression for the distribution of $Y = g(X)$ is available or when the probability distribution of X is not fully specified, the so-called *delta-method* provides approximate formulas for the mean and covariance of Y from the mean and covariance of X.

Let us consider the function $g : \mathbb{R}^m \mapsto \mathbb{R}^q$ and assume that g is differentiable at point μ_X. We have:

$$\begin{cases} \mathbb{E}\{g(X)\} \approx g(\mathbb{E}\{X\}) \\ \operatorname{cov}(g(X)) \approx J(\mu_X) \operatorname{cov}(X) J^T(\mu_X) \end{cases} \quad [1.49]$$

PROOF.– We denote $\mu_X = \mathbb{E}\{X\}$. Using the first-order Taylor expansion of g in the neighborhood of μ_X, we write:

$$Y = g(X) \approx g(\mu_X) + J(\mu_X)(X - \mu_X) \quad [1.50]$$

where

$$J(\mu_X) = \left.\frac{\partial y}{\partial x}\right|_{x=\mu_X} = \begin{bmatrix} \frac{\partial g_0}{\partial x_0}(\mu_X) & \cdots & \frac{\partial g_0}{\partial x_{m-1}}(\mu_X) \\ \vdots & & \vdots \\ \frac{\partial g_{q-1}}{\partial x_0}(\mu_X) & \cdots & \frac{\partial g_{q-1}}{\partial x_{m-1}}(\mu_X) \end{bmatrix}$$

is the $q \times m$ Jacobian matrix of g performed at point μ_X. Therefore, taking the expectation of [1.50], we get at first order

$$\mathbb{E}\{Y\} \approx g(\mu_X) + J(\mu_X) \times \mathbb{E}\{X - \mu_X\} = g(\mu_X) + 0$$

then

$$Y - \mathbb{E}\{Y\} \approx J(\mu_X)(X - \mu_X)$$

Therefore, according to the definition [1.21] of cov (Y), we have:

$$\operatorname{cov}(g(X)) \approx J(\mu_X)\operatorname{cov}(X)J^T(\mu_X)$$

It is worth noting that cov $(g(X))$ is a $q \times q$ matrix and cov (X) is a $m \times m$ matrix.

EXERCISE 1.2.– δ-method (see p. 61)

Consider two random variables $\{X_0, X_1\}$, Gaussian and independent, with means of μ_0 and μ_1 respectively, and with the same variance σ^2. Using the pair $\{X_0, X_1\}$, we consider the pair $\{R, \theta\}$ using the one-to-one mapping:

$$\{X_0, X_1\} = h(R, \theta) : \begin{cases} X_0 = R\cos(\theta) \in \mathbb{R} \\ X_1 = R\sin(\theta) \in \mathbb{R} \end{cases} \Leftrightarrow$$

$$\{R, \theta\} = g(X_0, X_1) : \begin{cases} R = |X_0 + jX_1| = \sqrt{X_0^2 + X_1^2} \in \mathbb{R}^+ \\ \theta = \arg(X_0 + jX_1) \in (0, 2\pi) \end{cases}$$

Use the δ-method to determine the covariance of the pair (R, θ). Use this result to deduce the variance of R. This may be compared with the theoretical value given by:

$$\operatorname{var}(R) = 2\sigma^2 + (\mu_0^2 + \mu_1^2) - \frac{\pi\sigma^2}{2}L_{1/2}^2\left(\frac{-(\mu_0^2 + \mu_1^2)}{2\sigma^2}\right)$$

where $L_{1/2}(x) = {}_1F_1\left(-\frac{1}{2}; 1; x\right)$ is the hypergeometric function (scipy.stats.hypergeom function in *Python*®). The later can be represented by an integral:

$${}_1F_1(a; b; x) = \frac{\Gamma(b)}{\Gamma(a)\Gamma(b-a)} \int_0^1 e^{xu} u^{a-1}(1-u)^{b-a-1} du$$

if $\operatorname{Re}(b) > \operatorname{Re}(a) > 0$, condition which is fulfilled here.

We see that, when $(\mu_0^2 + \mu_1^2)/\sigma^2$ tends toward infinity, var (R) tends toward σ^2. Additionally, when $\mu_0 = \mu_1 = 0$, we have var $(R) = (4-\pi)\sigma^2/2 \approx 0.43\sigma^2$.

1.6. Fundamental theorems of statistics

The law of large numbers and the central limit theorem form the basis of statistical methods, and are essential to the validity of Monte-Carlo methods, which are briefly presented in Chapter 4. The first theorem, often (erroneously) referred to as a *law*, states that the empirical mean converges in law to the statistical mean; the second theorem states that this convergence is "distributed in a Gaussian manner".

THEOREM 1.8 (Law of large numbers).– Let X_n be a series of random vectors of dimension d, independent and identically distributed, with a mean vector $m = \mathbb{E}\{X_0\} \in \mathbb{R}^d$ and finite covariance. In this case,

$$\frac{1}{N}\sum_{n=0}^{N-1} X_n \xrightarrow{a.s.}_{N\to+\infty} \mathbb{E}\{X_0\} = m$$

and convergence is almost sure (a.s.).

One fundamental example is that of empirical frequency, which converges toward the probability. Let X_n be a series of N random variables with values in $a_0, a_1, \ldots, a_{J-1}$ and let f_j be the empirical frequency, defined as the ratio between the number of values equal to a_j and the total number N. In this case:

$$f_j = \frac{1}{N}\sum_{n=0}^{N-1} \mathbb{1}(X_n = a_j) \xrightarrow{a.s.} \mathbb{E}\{\mathbb{1}(X_a = a_j)\} = \mathbb{P}\{X_0 = a_j\}$$

THEOREM 1.9 (Central limit theorem (CLT)).– Let X_n be a series of random vectors of dimension d, independent and identically distributed, of mean vector $m = \mathbb{E}\{X_0\}$ and covariance matrix $C = \text{cov}(X_0)$. In this case:

$$\sqrt{N}\left(\frac{1}{N}\sum_{n=0}^{N-1} X_n - m\right) \xrightarrow{d}_{N\to+\infty} \mathcal{N}(0, C)$$

with convergence in distribution.

Convergence in distribution is defined as follows:

DEFINITION 1.21 (Convergence in distribution).– A set of r.v. U_N is said to converge in distribution toward a r.v. U if, for any bounded continuous function f, when N tends toward infinity, we have:

$$\mathbb{E}\{f(U_N)\} \to_{N\to\infty} \mathbb{E}\{f(U)\} \qquad [1.51]$$

Theorem 1.9 is the basis for calculations of confidence intervals (see definition 2.7), and is used as follows: we approximate the probability distribution of the random vector $\sqrt{N}\left(N^{-1}\sum_{n=0}^{N-1} X_n - m\right)$, for which the expression is often impossible to calculate, by the Gaussian distribution.

EXAMPLE 1.1 (Application of the CLT).– For illustrative purposes, consider the Gaussian case where $d=1$, taking $\widehat{m}_N = N^{-1}\sum_{n=0}^{N-1} X_n$. Show that:

$$\mathbb{P}\left\{\widehat{m}_N - \frac{1.96\sigma}{\sqrt{N}} < m \leq \widehat{m}_N + \frac{1.96\sigma}{\sqrt{N}}\right\} \approx 0.95 \qquad [1.52]$$

HINTS: For any $c > 0$ and from theorem 1.9:

$$\mathbb{P}\left\{\sqrt{N}(\widehat{m}_N - m) \in (-\varepsilon, +\varepsilon)\right\} \approx 2\int_0^\varepsilon \frac{1}{\sigma\sqrt{2\pi}} e^{-u^2/2\sigma^2} du$$

Letting $\epsilon = c\sigma$, we can rewrite:

$$\mathbb{P}\left\{\widehat{m}_N - \frac{c\sigma}{\sqrt{N}} < m \leq \widehat{m}_N + \frac{c\sigma}{\sqrt{N}}\right\} \approx 2\int_0^c \frac{1}{\sqrt{2\pi}} e^{-t^2/2} dt$$

Aiming for a probability equal to 0.95, the Gaussian table yields $c = 1.96$, hence [1.52] (in *Python*® type `scipy.stats.norm.isf(0.05/2.0)`).

[1.52] means that there is 95% of chance that the true value of m belongs to the interval $(\widehat{m}_N - 1.96\sigma/\sqrt{N}, \widehat{m}_N + 1.96\sigma/\sqrt{N})$. As expected, the smaller the value of σ and/or the higher the value of N, the narrower the confidence interval.

In section 2.6, we will see the expression of the confidence interval when σ is unknown. ∎

EXERCISE 1.3 (Asymptotic confidence interval from the CLT (see p. 62)).–
Consider a sequence of N independent random Bernoulli variables X_k, such that $\mathbb{P}\{X_k = 1\} = p$. To estimate the proportion p, we consider $\widehat{p} = \frac{1}{N}\sum_{k=0}^{N-1} X_k$.

1) Using the central limit theorem 1.9, determine the asymptotic distribution of $\widehat{p}$.

2) Use the previous result to derive the approximate expression of the probability that p will lie within the interval between $\widehat{p} - \epsilon/\sqrt{N}$ and $\widehat{p} + \epsilon/\sqrt{N}$.

3) Use this result to deduce an interval which ensures that this probability will be higher than $100\,\alpha\%$, expressed as a function of N and $\alpha = 0.95$.

4) Write a program which verifies this asymptotic behavior.

The following theorem, known as the *continuity theorem*, allows us to extend the central limit theorem to more complicated functions:

THEOREM 1.10 (Continuity).– Let U_N be a series of random vectors of dimension d, such that:

$$\sqrt{N}(U_N - m) \xrightarrow{d}_{N \to +\infty} \mathcal{N}(0_d, C)$$

and let g be a function $\mathbb{R}^d \mapsto \mathbb{R}^q$ supposed to be twice continuously differentiable. Thus,

$$\sqrt{N}(g(U_N) - g(m)) \xrightarrow{d}_{N \to +\infty} \mathcal{N}(0_q, \Gamma)$$

where $\Gamma = \partial g(m) \, C \, \partial^T g(m)$ and where:

$$\partial g(u) = \begin{bmatrix} \dfrac{\partial g_0(u_0, \ldots, u_{d-1})}{\partial u_0} & \cdots & \dfrac{\partial g_0(u_0, \ldots, u_{d-1})}{\partial u_{d-1}} \\ \vdots & & \vdots \\ \dfrac{\partial g_q(u_0, \ldots, u_{d-1})}{\partial u_0} & \cdots & \dfrac{\partial g_q(u_0, \ldots, u_{d-1})}{\partial u_{d-1}} \end{bmatrix}$$

Applying theorem [1.10], consider the function associating vector U_N with its ℓ-th component, which is written as:

$$U_N \mapsto U_{N,\ell} = E_\ell^T U_N$$

where E_ℓ is the vector of dimension d, all components of which are equal to 0, with the exception of the ℓ-th, equal to 1. Direct application of the theorem gives:

$$\sqrt{N}(U_{N,\ell} - m_\ell) \xrightarrow{d} \mathcal{N}(0, C_{\ell\ell})$$

where m_ℓ is the ℓ-th component of m and $C_{\ell\ell}$ is the ℓ-th diagonal element of C.

1.7. A few probability distributions

This section presents a non-exhaustive list of certain other important probability distributions.

Binomial distribution: noted $\mathcal{B}(N, p)$ defined by:

$$\mathbb{P}\{X = n\} = \begin{cases} C_N^n p^n (1-p)^{N-n} & \text{if } n = 0, \ldots, N \\ 0 & \text{if not} \end{cases} \quad [1.53]$$

Poisson distribution: noted $\mathcal{P}(\lambda)$ defined for $n \in \mathbb{N}$ by:

$$\mathbb{P}\{X = n\} = \frac{\lambda^n}{n!}e^{-\lambda} \qquad [1.54]$$

Uniform distribution over (a, b): noted $\mathcal{U}(a, b)$ of density

$$p_X(x; a, b) = \frac{1}{b-a}\mathbb{1}(x \in (a, b)) \qquad [1.55]$$

where $a < b$. The mean is equal to $(b + a)/2$ and the variance is equal to $(b-a)^2/12$.

Exponential distribution: noted $E(\theta)$, of density

$$p_X(x; \theta) = \theta^{-1}e^{-x/\theta}\mathbb{1}(x \geq 0) \qquad [1.56]$$

with $\theta > 0$. The mean is equal to θ and the variance to θ^2. We can easily demonstrate that $E(\theta) = \theta E(1)$.

Cauchy distribution: noted $\mathcal{C}(x_0, a)$, of density

$$p_X(x; x_0, a) = \frac{1}{\pi a \left(1 + \left(\frac{x - x_0}{a}\right)^2\right)} \qquad [1.57]$$

with $x_0 \in \mathbb{R}$ and $a > 0$. The Cauchy distribution has no finite moments.

Rayleigh distribution: of density

$$p_X(x; \sigma^2) = \frac{x}{\sigma^2}e^{-x^2/2\sigma^2}\mathbb{1}(x \geq 0) \qquad [1.58]$$

with $\sigma > 0$. The mean is equal to $\sigma\sqrt{\pi/2}$ and the variance to $\sigma^2(4 - \pi)/2$, see exercise 1.1.

Gamma distribution: noted $G(k, \theta)$, of density

$$p_X(x; (k, \theta)) = \frac{1}{\Gamma(k)\theta^k}e^{-x/\theta}x^{k-1}\mathbb{1}(x > 0) \qquad [1.59]$$

where $\theta \in \mathbb{R}^+$ and $k \in \mathbb{R}^+$. The mean is equal to $k\theta$ and the variance to $k\theta^2$. Note that $E(\theta) = G(1, \theta)$.

ki square distribution with k d.o.f.: the r.v. $Y = \sum_{i=0}^{k-1} X_i^2$, where X_i are k Gaussian, independent, centered r.v.s of variance 1 follows a χ^2 (pronounced "ki square") distribution with k degrees of freedom (d.o.f.). The mean is equal to k and the variance to $2k$.

Fisher distribution with (k_0, k_1) *d.o.f.*: noted $F(k_0, k_1)$. Let X and Y be two real, centered Gaussian vectors of respective dimensions k_0 and k_1, with respective covariance matrices I_{k_0} and I_{k_1}, and independent of each other, then the r.v.

$$F_{k_0,k_1} = \frac{k_0^{-1} X^T X}{k_1^{-1} Y^T Y} \qquad [1.60]$$

follows a Fisher distribution with (k_0, k_1) d.o.f.

Student distribution with k d.o.f.: noted T_k. Let X be a real, centered Gaussian vector, with a covariance matrix I_k, and Y a real, centered Gaussian vector, of variance 1 and independent of X. The r.v.

$$T_k = \frac{Y}{\sqrt{k^{-1} \sum_{i=0}^{k-1} X_i^2}} \qquad [1.61]$$

follows a Student distribution with k d.o.f.

We can show that if Z follows a Student distribution with k degrees of freedom, then Z^2 follows a Fisher distribution with $(1, k)$ degrees of freedom.

EXAMPLE 1.2 (A few distributions).– The following program displays the shapes of a few distributions and the histograms of the *Python*® random generators, performed on many samples:

```
# -*- coding: utf-8 -*-
"""
Created on Wed Jun  8 09:22:45 2016
****** afewdistributions
@author: maurice
"""
from numpy import exp
from scipy.stats import norm, gamma, f, t, chi2
from numpy.random import gamma as rndgamma, f as rndf, randn, 
            exponential
from numpy.random import chisquare as rndchi2, standard_t as rndt
from matplotlib import pyplot as plt
N = 10000; xG=randn(N); bE=0.2; xE=exponential(scale=bE,size=N)
thetag = 2; kg=3; xgamma = rndgamma(shape=kg,scale=thetag,size=N)
dofF = (30,40); xF = rndf(dofF[0],dofF[1],size=N)
doft = 10; xt = rndt(df=doft, size=N)
dofchi2 = 20; xchi2 = rndchi2(df=dofchi2, size=N)
#===== display
```

```
bins = 50; plt.clf()
plt.subplot(321); histG = plt.hist(xG,normed='True', bins=bins)
plt.plot(histG[1], norm.pdf(-histG[1],0,1),'.-r')
plt.yticks([]); plt.xticks(fontsize=8)
plt.title('standard Gaussian', fontsize=8)
plt.subplot(322); histE = plt.hist(xE,normed='True', bins=bins)
plt.plot(histE[1], exp(-histE[1]/bE)/bE,'.-r')
plt.yticks([]); plt.xticks(fontsize=8)
plt.title('Exponential with parameter %4.2f'%bE, fontsize=8)
plt.subplot(323);histgamma=plt.hist(xgamma,normed='True',bins=bins)
plt.plot(histgamma[1],gamma.pdf(histgamma[1],kg,scale=thetag),'.-r')
plt.yticks([]); plt.xticks(fontsize=8)
plt.title('Gamma k=%i, theta=%4.2f'%(kg,thetag), fontsize=8)
plt.subplot(324); histF = plt.hist(xF,normed='True', bins=bins)
plt.plot(histF[1], f.pdf(histF[1],dofF[0],dofF[1]),'.-r')
plt.yticks([]); plt.xticks(fontsize=8)
plt.title('Fisher with (%i,%i) d.o.f'%dofF, fontsize=8)
plt.subplot(325); histt = plt.hist(xt,normed='True', bins=bins)
plt.plot(histt[1], t.pdf(histt[1],doft),'.-r')
plt.yticks([]); plt.xticks(fontsize=8)
plt.title('Student with %i d.o.f'%doft, fontsize=8)
plt.subplot(326);histchi2=plt.hist(xchi2,normed='True',bins=bins)
plt.plot(histchi2[1], chi2.pdf(histchi2[1],dofchi2),'.-r')
plt.yticks([]); plt.xticks(fontsize=8)
plt.title('$\chi^2$ with %i d.o.f'%dofchi2, fontsize=8)
```

2

Statistical Inferences

2.1. First step: visualizing data

In data analysis, a picture is often better than a thousand words. When we start analyzing data, the first step is not to run a complex statistical tool, but should be to visualize the data in a graph. This will allow us to understand the basic nature of the data. To meet this requirement, the following techniques are generally considered: *scatter plot, histogram, boxplot* and *Q-Q plot*.

2.1.1. Scatter plot

The scatter plot is a plot in which Cartesian coordinates are used to display values of pairs of variables. A scatter plot can suggest various kinds of correlations and trends between two variables. Typically, the scatter plot is commonly used to determine whether the relationship between variables is linear or not. The following program displays the scatter plot of 6 variables of the exogenous variables of the dataset statsmodels.api.datasets.star98:

```
# -*- coding: utf-8 -*-
"""
Created on Sun Jul  3 07:53:57 2016
****** scatterplotexample
@author: maurice
"""
import statsmodels.api as sm
from matplotlib import pyplot as plt
data = sm.datasets.star98.load()
P = 6; C=int((P-1)*P/2);
DP = 13; X0 = data.exog[:,DP:DP+P]
```

```
plt.figure(1); plt.clf(); cp=0;
for i1 in range(P):
    for i2 in range(i1+1,P):
        cp=cp+1;
        plt.subplot(4,4,cp);
        plt.plot(X0[:,i1],X0[:,i2],'.')
        plt.xticks([]); plt.yticks([])
        plt.title('(%i,%i)'%(i1,i2), fontsize=10)
plt.subplot(4,4,16)
for ip in range(P):
    nametext=('%i: %s'%(ip,data.names[DP+ip]))
    plt.text(0,0.9-ip/P,nametext,fontsize=10)
    plt.xticks([]); plt.yticks([])
    plt.box('off')
plt.show()
```

Scatter plots are shown in Figure 2.1. We see that the couples $(0,5)$, on the one hand, and $(3,4)$, on the other hand, present a linear trend. Moreover, the couple $(3,4)$ seems to present a variance that increases with index number. Also, the scatter plot of the couple $(1,2)$ suggests a possible linear trend. Therefore, it is useful to define quantitatively to assess the level of confidence when we decide that the trend is linear.

Sometimes a nonlinear trend can appear which can be easily transformed into a linear trend using a simple transformation such as the log function.

We can also observe periodic trends that are often related to annual or daily periodicities. Such a trend is said to be seasonal. For example, Figure 2.2 shows the annual trend and may also be a parabolic trend (see exercise 2.11).

2.1.2. *Histogram/boxplot*

A histogram is a graphical representation of the distribution of numerical values. The abscissa corresponds to the range of values and the ordinate to the numbers of values in a set of consecutive disjoint intervals. In section 2.6.7, more details will be provided on the relationship between the histogram and the probability distribution.

A boxplot determines the sequence of consecutive disjoint intervals, each containing a given percentage of values, for example the four intervals containing 25% of the values.

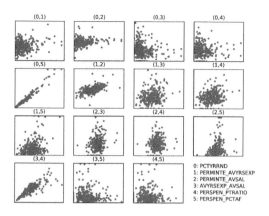

Figure 2.1. Scatter plots of all the pairs of six variables. The couples $(0,5)$ and $(3,4)$ present a clear linear trend

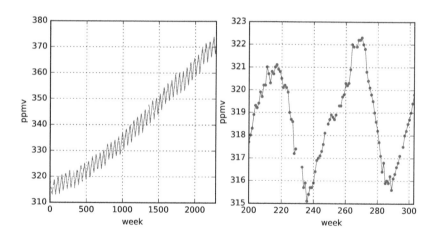

Figure 2.2. Atmospheric CO_2 concentration from continuous air samples at Mauna Loa Observatory, Hawaii. The recorded data are weekly averages of parts per million by volume (ppmv), observed from March 1958 to December 2001. On the RHS, a zoom on 104 weeks shows a clear annual trend. Some data are missing

Type and run the following program:

```
# -*- coding: utf-8 -*-
"""
Created on Tue Aug 23 08:08:15 2016
```

```
****** histograminbrief
@author: maurice
"""

from matplotlib import pyplot as plt
import sklearn.datasets as sk
diabetes = sk.load_diabetes(); data = diabetes['data']; ip = 2;
fighist = plt.figure(num=1,figsize=(6,4)); plt.clf();
plt.subplot(211); plt.hist(data[:,ip], bins=10);
plt.subplot(212); plt.boxplot(data[:,ip],vert=0); plt.show()
```

The top graph of Figure 2.3 shows the histogram with 10 equal intervals (nbins=10 in the program) distributed between the minimal and maximal values. The observed shape does not suggest a Gaussian distribution. The bottom graph shows the boxplot. The central box consists of 50% of the values and the vertical line indicates the empirical median which is the value M, such that 50% of the set of values is under M. (see definition 2.11).

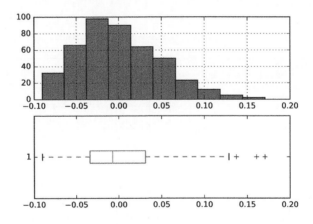

Figure 2.3. *Top: the histogram consists of 10 bins of equal size. The ordinates correspond to the number of values in different disjoint intervals. Bottom: the boxplot consists of a central box containing 50% of the values. The vertical line indicates the empirical median. The plus symbols indicate outliers.*

2.1.3. *Q-Q plot*

A Q-Q plot is a graphical method for testing a theoretical distribution with observations. It is used to plot the theoretical quantiles (see expression [2.5]) as a function of the empirical quantiles, hence its name. For a given distribution $\mathbb{P}$, the quantile associated with the percent value p is the value x, such that $\mathbb{P}\{X \leq x\} = 1 - p$. It is the inverse function of the cumulative function. For more details on the quantile estimate, see exercise 2.35. Here, we only need to interpret the results.

In *Python®*, the function scipy.stats.probplot returns, for a given distribution specified by the argument dist, (i) the theoretical quantiles as a function of the empirical quantiles, sorted in increasing order, and (ii) the closest straight line characterized by its slope and its origin ordinate (also called the intercept) and the root square of the *coefficient of determination* denoted by R^2, see section 2.6.3.10. The closer the R^2 to 1, the closer the Q-Q plot to the straight line and the better the fit with the tested distribution.

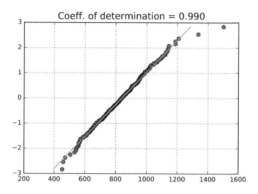

Figure 2.4. *Q-Q plot of the math grades of 303 students. The theoretical quantile values are derived from a Gaussian distribution. The graph indicates that the data can be considered as Gaussian*

The file statsmodels.api.datasets.star98 consists of the math scores of 303 students. The following program applies a Q-Q plot to these data. The results, shown in Figure 2.4, indicate that the distribution is likely to be Gaussian.

```
# -*- coding: utf-8 -*-
"""
Created on Sun Jul  3 10:49:09 2016
****** qqplotexample
@author: maurice
"""
import statsmodels.api as sm
from numpy import size
from scipy.stats import probplot
```

```
from matplotlib import pyplot as plt
data = sm.datasets.star98.load()
i0 = 14; X0 = data.exog[:,i0]; N = size(X0,0)
resultspp = probplot(X0, dist="norm")
sortX0 = resultspp[0][1]; qq = resultspp[0][0]
slope = resultspp[1][0]; intercept = resultspp[1][1]
R2 = resultspp[1][2]**2; plt.figure(2); plt.clf()
plt.plot(sortX0,qq,'ob'); plt.hold('on')
plt.plot(intercept+(slope*qq[0],slope*qq[N-1]),(qq[0],qq[N-1]),'r')
plt.hold('off'); plt.grid('on');
plt.title('Coeff. of determination = %4.4e'%R2);plt.show()
```

2.2. Reduction of dataset dimensionality

The dimension of the data is the number of variables that are measured on each observation. One of the main issues is that, in many cases, not all the variables are of interest for understanding the underlying phenomena. Therefore, it is still useful in many applications to reduce the dimension of the original dataset prior to any modeling of the data.

In this section, we only consider one approach, namely principal components analysis (PCA), to reduce the dimensionality of a complex dataset of observations. PCA consists of projecting onto a space of a small dimension while preserving as much as possible the useful variability.

We also present the linear discriminant analysis (LDA) that is closely related to the PCA but attempts to model the difference between different classes of data. The LDA is commonly used for dimensionality reduction before later classification.

The two methods have the advantage of being linear and make no precise statistical assumptions concerning the data distribution. They are widely used in a number of applications.

2.2.1. *PCA*

Consider N observations $x_0, \ldots, x_{N-1}$, each one of length d associated with a set of d features. We let:

$$X = \begin{bmatrix} x_{0,0} & x_{0,1} & \cdots & x_{0,N-1} \\ \vdots & \ddots & & \vdots \\ x_{d-1,0} & x_{d-1,1} & \cdots & x_{d-1,N-1} \end{bmatrix} = \begin{bmatrix} x_0 & \cdots & x_{N-1} \end{bmatrix}$$

X can be understood from two points of view:

– either as a set of N columns, each one of them representing the *features* associated with the same individual. This leads to N points in the space $\mathbb{R}^d$;

– or as a set of d rows, each one of them representing the same factor for N individuals. This leads to d points in the space $\mathbb{R}^N$.

Here we will reduce the number of factors to keep only the k most significant ones, eventually by linearly combining some of them.

First, we consider the case where $k = 1$. We have to search in $\mathbb{R}^d$ the direction of a unit vector v such that the projection of the set of the x_n onto this direction leads to the scatter of N points with the highest dispersion.

This can be interpreted as follows: if we have to keep only one component, it might also be the one that best separates all of the points. A classic criterion to evaluate this dispersion is to consider the sum of the distances between all of the projected points. Note that the projection of x_n onto v is given by $vv^T x_n$ and that the distance between any two points is written as $\|vv^T x_i - vv^T x_j\|^2 = v^T(x_i - x_j)(x_i - x_j)^T v$. The criterion to be maximized is then written as:

$$J(v) = \frac{1}{2N^2} v^T \left(\sum_{i=0}^{N-1} \sum_{j=0}^{N-1} (x_i - x_j)(x_i - x_j)^T \right) v$$

Let $\bar{x} = N^{-1} \sum_{j=0}^{N-1} x_j \in \mathbb{R}^d$.

Therefore:

$$J(v) = \frac{1}{2N^2} v^T \left(\sum_{i=0}^{N-1} \sum_{j=0}^{N-1} [(x_i - \bar{x}) - (x_j - \bar{x})][(x_i - \bar{x}) - (x_j - \bar{x})]^T \right) v$$

$$= v^T R_N v$$

where we let:

$$R_N = \frac{1}{N} \sum_{n=0}^{N-1} (x_n - \bar{x})(x_n - \bar{x})^T \qquad [2.1]$$

Note that the $(d \times d)$ matrix R_N can be interpreted as a covariance matrix. Hence, the problem can be laid out as follows:

$$\begin{cases} \max_v J(v) \\ \text{with} \quad v^T v - 1 = 0 \end{cases}$$

The Lagrange multiplier method leads us to the following equivalent problem:

$$\begin{cases} \max_v \left(v^T R_N v - \lambda(v^T v - 1) \right) \\ v^T v - 1 = 0 \end{cases}$$

Setting the gradient to zero with respect to v, we have:

$$\begin{cases} R_N v - \lambda v = 0 \\ v^T v - 1 = 0 \end{cases}$$

The first equation means that v is an eigenvector of the matrix R_N. If λ denotes the eigenvalue associated with v, then $J(v) = \lambda \geq 0$. Hence, the maximum is reached when v is chosen as the eigenvector of R_N associated with the highest eigenvalue, hence the name of this method, *Principal Component Analysis* (or *PCA*).

By calculating the product $v^T X$, we get a vector of $\mathbb{R}^N$ which can be interpreted as the "mean" factor that best characterizes, by itself, the N individuals, resulting in a good separation. However, this "factor" does not really exist; it is merely a linear combination of factors that are actually observed.

The previous result can easily be generalized: the k principal directions are the k eigendirections of the *highest* eigenvalues. This leads to the following algorithm:

Data: X array of size $d \times N$, $k < d$
Result: Z array of size $k \times N$, V array of size $d \times k$
begin
    Compute the covariance matrix R as [2.1];
    Compute the eigendecomposition $R = UDU^T$;
    Sort the eigenvalues in increasing order;
    Selecting $V = U_{d-k:d-1}$, compute $Z = V^T X$;
end

Algorithm 1: PCA algorithm

2.2.2. LDA

The LDA is a method to find a linear combination of features that allows us to separate two or more classes of objects or events. Unlike the PCA, the LDA is a supervised method, meaning that the response is known during the learning phase. However, as with the PCA, the LDA is commonly used for dimensionality reduction before later classification.

We now consider g groups of individuals, where each group is comprised of N_ℓ individuals for which d factors are measured. The data can then be represented in the form of g matrices of the type:

$$X_\ell = \begin{bmatrix} x_{0,0} & x_{0,1} & \cdots & x_{0,N_\ell-1} \\ \vdots & \ddots & & \vdots \\ x_{d-1,0} & x_{d-1,1} & \cdots & x_{d-1,N_\ell-1} \end{bmatrix} = \begin{bmatrix} x_{\ell,0} & \cdots & x_{\ell,N_\ell-1} \end{bmatrix}$$

with ℓ ranging from 0 to $g-1$.

In the space $\mathbb{R}^d$, we obtain g scatters containing, respectively, $N_1, \ldots, N_g$ points. We let $N = \sum_{\ell=0}^{g-1} N_\ell$.

The goal of the *Linear Discriminant Analysis* (or *LDA*) is to find the best separation for these g scatters of points. To achieve this, we must first introduce the following definitions:

– the mean, or barycenter, of a group:

$$m_\ell = \frac{1}{N_\ell} \sum_{j=0}^{N_\ell-1} x_{\ell,j}$$

– the overall mean of the g groups:

$$m = \frac{1}{N} \sum_{\ell=0}^{g-1} \sum_{j=0}^{N_\ell-1} x_{\ell,j} = \frac{\sum_{\ell=0}^{g-1} N_\ell m_\ell}{\sum_{\ell=0}^{g-1} N_\ell} \qquad [2.2]$$

– the intraclass covariance (internal to the considered class) defined by:

$$R_I = \frac{\sum_{\ell=0}^{g-1} N_\ell R_\ell}{\sum_{\ell=0}^{g-1} N_\ell} \text{ with } R_\ell = \frac{1}{N_\ell} \sum_{j=0}^{N_\ell-1} (x_{\ell,j} - m_\ell)(x_{\ell,j} - m_\ell)^T$$

which leads us to:

$$R_I = \frac{1}{N} \sum_{\ell=0}^{g-1} \sum_{j=0}^{N_\ell-1} x_{\ell,j} x_{\ell,j}^T - \frac{1}{N} \sum_{\ell=0}^{g-1} N_\ell m_\ell m_\ell^T \qquad [2.3]$$

which can be interpreted as the mean of the dispersions inside of each group.

– the extraclass covariance defined by:

$$R_E = \frac{1}{N} \sum_{\ell=0}^{g-1} N_\ell (m_\ell - m)(m_\ell - m)^T \qquad [2.4]$$

which can be interpreted as the dispersion of the barycenters of each class with respect to the overall mean.

– the total covariance defined by:

$$R = \frac{1}{N} \sum_{\ell=0}^{g-1} \sum_{j=0}^{N_\ell-1} (x_{\ell,j} - m)(x_{\ell,j} - m)^T = \frac{1}{N} \sum_{\ell=0}^{g-1} \sum_{j=0}^{N_\ell-1} x_{\ell,j} x_{\ell,j}^T - mm^T$$

These covariance matrices are $(d \times d)$ matrices. It can easily be proved that:

$$R = R_I + R_E$$

PROPERTY 2.1.– R_E is of rank $r = \min(d, g-1)$.

PROOF.– Indeed,

$$NR_E = \sum_{\ell=0}^{g-1} N_\ell (m_\ell - m)(m_\ell - m)^T$$

where m is given by expression [2.2]. We let:

$$M = \left[\sqrt{N_0}(m_0 - m) \ \cdots \ \sqrt{N_{g-1}}(m_{g-1} - m) \right]$$

The $d \times g$ matrix M verifies $MM^T = \sum_{\ell=0}^{g-1} N_\ell (m_\ell - m)(m_\ell - m)^T = NR_E$. On the other hand,

$$M \begin{bmatrix} \sqrt{N_0} \\ \vdots \\ \sqrt{N_{g-1}} \end{bmatrix} = \sum_{j=0}^{g-1} N_j m_j - mN = 0$$

Therefore, the rank of M is at most equal to $\min(d, g-1)$. This means that the space spanned by the columns of M is at most of dimension $\min(d, g-1)$. This is also true for $NR_E = MM^T$. ∎

From property 2.1, it follows that, if $g - 1 < d$, R_E is not invertible.

We will now find the direction carried by the vector v, such that the intraclass dispersion is minimal and the interclass dispersion is maximal: graphically speaking, the scatters are farther away from each other and more compact. To achieve this objective, one possible criterion is to *minimize* the evaluation function defined by:

$$J(v) = \frac{v^T R_I v}{v^T R_E v}$$

This amounts to searching v, such that:

$$\min_v v^T R_I v \quad \text{with} \quad v^T R_E v = 1 \qquad [2.5]$$

Using the Lagrange multiplier method, we end up with the following equivalent problem:

$$\begin{cases} \min_v \left(v^T R_I v + \frac{1}{\lambda}(1 - v^T R_E v)\right) \\ \text{with} \quad 1 - v^T R_E v = 0 \end{cases}$$

By setting the gradient to zero with respect to v, we find that v is given by:

$$R_E v = \lambda R_I v \qquad [2.6]$$

Note that if v is a solution of equation [2.6], then $J(v) = 1/\lambda$ with $\lambda \geq 0$. Now we have to solve the equation [2.6] with respect to v and choose v associated with the maximal value of λ. The resolution of [2.6] is a well-known problem, which is called the *generalized eigenvalue problem*. In our case, R_E and R_I are positive; therefore, there exists a basis of generalized eigenvectors that are the solution of [2.6]. In *Python*®, the resolution of [2.6] is obtained with the function eig(RE,RI).

Let us note that the eigenvectors of the classical eigendecomposition, equation $Rv = \lambda v$, are orthogonal, while eigenvectors of the generalized eigendecomposition, equation $R_E v = \lambda R_I v$, are not.

To increase the capability of classification, we can choose more than one eigenvector verifying equation [2.6]. Let v_{d-k}, ..., v_{d-1} be the k eigenvectors associated with the k *greatest* generalized eigenvalues of equation [2.6]. By compiling these k vectors, we get the $(d \times k)$ matrix:

$$V = \begin{bmatrix} v_{d-k} \end{bmatrix} \qquad [2.7]$$

This means that for each of the g families, the vectors are given by:

$$y_{\ell,j} = V^T x_{\ell,j} \qquad [2.8]$$

which give the representative points in $\mathbb{R}^k$ of each image. Theoretically, each of the g scatter has a minimal dispersion, and all of the scatters are as far away from each other as possible. This leads to the following algorithm:

Data: X array of size $d \times N$, y array of size N, $k < d$
Result: Z array of size $k \times N$, V array of size $d \times k$
begin
    Compute the matrix R_I as [2.3];
    Compute the matrix R_E as [2.4];
    Compute the generalized eigen-decomposition of (R_I, R_E) that gives U, D;
    Sort the d eigenvalues of D in increasing order;
    Selecting $V = U_{d-k:d-1}$, compute $Z = V^T X$
end

Algorithm 2: LDA algorithm

EXERCISE 2.1 (Dimensionality reduction on the iris dataset).– (see 63)

The file located at `sklearn.load_iris()` consists of 3 classes of iris: Setosa, Versicolour and Virginica. Each class consists of 50 examples, each example described by 4 features.

Write a program reducing the dimension from 4 to 2, (i) firstly with PCA approach i.e. without considering the iris classes and (ii) secondly with LDA approach taking into account the iris classes.

For each example we obtain with this method a 2D feature that can be considered as a point in the 2D plane. Plot these points for the 3 classes. Conclude.

These results will be used in exercise 2.20 for classification.

2.3. Some vocabulary

Observations may be either quantitative, or qualitative with an order notion, or qualitative with no order notion. *Quantitative* values belong to $\mathbb{R}^n$. The values can be univariate ($n = 1$), bivariate ($n = 2$) or multivariate ($n > 2$). *Unordered qualitative* values include, for example, gender (man, woman) or place of residence. *Ordered qualitative* values include size (very tall, tall, small, very small), customer satisfaction level (good, bad) or opinion score (excellent, good, fair, poor, bad).

2.3.1. *Statistical inference*

The aim of *statistical inference* is to obtain conclusions based on observations modeled using random variables or vectors. A number of examples of statistical inference problems are as follows:

– estimating the value of a parameter, for example the mean;

– estimating an interval that has a 95% probability of containing the true parameter value;

– testing the hypothesis that a parameter belongs to a given region, for example the mean is greater than a given value;

– conducting regression analysis that addresses statistical techniques for estimating the relationships among variables;

– classifying and/or ranking observations.

We do not aim to provide exhaustive coverage in this discussion, but simply provide some examples of hypothesis testing, estimation and regression analysis.

In regression analysis, two sequences of values (X_n, Y_n) are usually observed. Observations X_n are considered as predictors or explanatory or independent variables[1], and observations Y_n as responses or explained or dependent variables. See, for example, the iris categories given in exercise 2.1 or the logistic regression presented in section 2.6.6.

In the case where response Y_n is quantitative, we speak of *regression*. When Y_n is qualitative without order, we speak of *classification*. Finally, when Y_n is qualitative and ordered, we speak of *ranking*.

2.4. Statistical model

In what follows, an element of the sample set $\mathcal{X}$ is denoted by x and the associated event set is denoted by $\mathcal{B}_\mathcal{X}$. The pair $\{\mathcal{X}, \mathcal{B}_\mathcal{X}\}$ is known as a measurable set [RUD 86]. Often, in the following, the sample set will be $\mathbb{R}^n$ and the associated event set will be the Borel σ-algebra derived from the natural open topology.

DEFINITION 2.1 (Statistical model).– A statistical model is a family of probability measures defined over the same measurable set $\{\mathcal{X}, \mathcal{B}_\mathcal{X}\}$ and indexed by $\theta \in \Theta$, which is written as:

$$\{\mathbb{P}_\theta\,; \theta \in \Theta\} \qquad [2.9]$$

When the set $\Theta \subset \mathbb{R}^p$ is of finite dimensions, the model is said to be *parametric*; otherwise, it is said to be *non-parametric*.

---

[1] Let us note that the term independent should not be confused with independent in the probability context.

In what follows, we shall mainly focus on parametric models, where $\mathcal{X} \in \mathbb{R}^n$, for which the probability distribution has either a discrete form or has a density with respect to the Lebesgue measure. This model is denoted as:

$$\{p(x; \theta); \theta \in \Theta\} \qquad [2.10]$$

θ is said to be *identifiable* if and only if:

$$\theta_0 \neq \theta_1 \Leftrightarrow p(x; \theta_0) \neq p(x; \theta_1)$$

We shall only consider models with identifiable parameters. It is essential that two different parameter values will not, statistically, produce the "same" observations.

2.4.1. *Notation*

In the following, $\mathbb{E}_\theta \{f(X)\}$ denotes the expectation of $f(X)$ when the distribution associated with the parameter value θ is used:

$$\theta \in \Theta \mapsto \mathbb{E}_\theta \{f(X)\} = \int_{\mathcal{X}} f(x) p(x; \theta) dx$$

EXAMPLE 2.1 (Gaussian model).– The probability measure P_θ is Gaussian over $\mathbb{R}$ with mean m and variance σ^2. Thus, the parameter $\theta = (m, \sigma^2) \in \Theta = \mathbb{R} \times \mathbb{R}^+$ and the model is parametric. If $\sigma \neq 0$, the probability has a density which is written as:

$$p(x; \theta) = \frac{1}{\sigma \sqrt{2\pi}} e^{-(x-m)^2 / 2\sigma^2}$$

In many applications, the statistical model is connected to a sequence of observations. For example, consider the observations modeled by n independent Gaussian random variables, with the same mean $m \in \mathbb{R}$ and the same variance $\sigma^2 > 0$. This model, which forms the basis for a significant number of important results, will be denoted as:

$$\{\text{ i.i.d. } \mathcal{N}(n; m, \sigma^2)\} = \ldots \qquad [2.11]$$

$$\left\{ p_X(x_0, \ldots, x_{n-1}; \theta) = \prod_{k=0}^{n-1} \frac{1}{\sigma \sqrt{2\pi}} e^{-(x_k - m)^2 / 2\sigma^2} \right\}$$

where parameter $\theta = (m, \sigma^2) \in \Theta = \mathbb{R} \times \mathbb{R}^+$.

Expression [2.11] may be generalized in cases where the observation is modeled by a sequence of n random vectors X_n of dimension d that are Gaussian and independent:

$$\{ \text{ i.i.d. } \mathcal{N}(n; m, C) \} = \ldots \qquad [2.12]$$

$$\left\{ p_X(x;\theta) = \prod_{k=0}^{n-1} \frac{1}{(2\pi)^{d/2} \sqrt{\det\{C\}}} e^{-\frac{1}{2}(x_k-m)^T C^{-1}(x_k-m)} \right\}$$

where $x_k = (x_{0,k}, \ldots, x_{d-1,k})$. Parameter $\theta = (m, C) \in \mathbb{R}^d \times \mathcal{M}_d^+$, where $\mathcal{M}_d^+$ denotes the set of positive square matrices of dimension d.

DEFINITION 2.2 (Likelihood).– Let us consider a parametric model defined by its probability density family $p_X(x; \theta)$. The likelihood is the function of Θ in $\mathbb{R}^+$ defined by:

$$\theta \mapsto p(X; \theta) \qquad [2.13]$$

Its logarithm is called the log-likelihood.

Likelihood plays a fundamental role in statistics.

DEFINITION 2.3 (Statistic/estimator).– A statistic or estimator is any measurable function of observations.

The statistical model $\{$ i.i.d. $\mathcal{N}(n; m, \sigma^2) \}$, where σ^2 is assumed to be known, is fundamentally different from model $\{$ i.i.d. $\mathcal{N}(n; m, \sigma^2) \}$, where σ^2 is an unknown parameter. In the first case, an estimator may contain the variable σ^2, whereas in the second case, it cannot.

2.5. Hypothesis testing

DEFINITION 2.4 (Hypothesis).– A hypothesis is a non-empty subset of Θ, which is said to be *simple* if it reduces to a singleton; in all other cases, it is said to be *composite*.

EXAMPLE 2.2 (Hypothesis simple/composite).– Consider the following model:

– $\{$ i.i.d. $\mathcal{N}(n; m, 1) \}$ of mean $m \in \Theta = \mathbb{R}$ and variance 1. The hypothesis $H_0 = \{0\}$ is simple, while the alternative hypothesis $H_1 = \Theta - H_0$ is composite;

– $\{$ i.i.d. $\mathcal{N}(n; m, \sigma^2) \}$ with $(m, \sigma^2) \in \Theta = \mathbb{R} \times \mathbb{R}^+$. The two hypotheses $H_0 = \{0\} \times \mathbb{R}^+$ and $H_1 = \Theta - H_0$ are composite.

A test of the hypothesis H_0 consists of defining a subset $\mathcal{X}_1 \subset \mathcal{X}$, such that if realization x belongs to $\mathcal{X}_1$, then H_0 is rejected. This is equivalent to defining the statistic $T(X) = \mathbb{1}(X \in \mathcal{X}_1)$ by taking values of 0 or 1 and such that:

$$\text{if } T(X) = \begin{cases} 0 & \text{then } H_0 \text{ is accepted,} \\ 1 & \text{then } H_0 \text{ is rejected.} \end{cases} \qquad [2.14]$$

The subset $\mathcal{X}_1$ is known as the *critical region* and the statistic $T(X)$ as the *critical test function*.

In some cases, a number η and a real-valued statistic $S(X)$ exist, such that the critical function may be written in the form $T(X) = \mathbb{1}(S(X) \geq \eta)$. The test is said to be *monolateral*. In some other cases, two numbers $\eta_1 < \eta_2$ and a real-valued statistic $S(X)$ exist, such that $T(X) = \mathbb{1}(S(X) \notin [\eta_1, \eta_2])$. In these cases, the test is said to be *bilateral*. $S(X)$ is referred to as *statistic of test*.

DEFINITION 2.5 (Significance level).– A test associated with the critical function $T(X)$ is said to have a significance level α if

$$\max_{\theta \in H_0} \mathbb{E}_\theta \{T(X)\} = \alpha \qquad [2.15]$$

Note that, since the random variable $T(X)$ takes its values in the set $\{0, 1\}$, $\mathbb{E}_\theta \{T(X)\} = \mathbb{P}_\theta \{T(X) = 1\}$. Thus, the significance level represents the probability of accepting H_1 when H_0 is true. It is also known as the *false alarm probability*.

DEFINITION 2.6.– A test of the hypothesis H_0 against the alternative $H_1 = \Theta - H_0$, associated with the critical function $T(X)$, is said to be uniformly most powerful (UMP) at the level α if, for any $\theta \in H_1$, its power $\mathbb{E}_\theta \{T(X)\}$ is higher than that of any other test at the level α. This is written as:

$$\exists\, T(X) \in \mathcal{T}_{H_0}(\alpha), \text{ s.t. } \forall\, S(X) \in \mathcal{T}_{H_0}(\alpha) \text{ and } \forall\, \theta \in H_1:$$
$$\mathbb{E}_\theta \{T(X)\} \geq \mathbb{E}_\theta \{S(X)\} \qquad [2.16]$$

where $\mathcal{T}_{H_0}(\alpha)$ denotes the set of tests of the hypothesis H_0 at the level α.

The power is interpreted as the probability of *detection*, which consists of accepting H_1 when H_1 is true.

Unfortunately, in most situations, there is no UMP test.

2.5.1. *Simple hypotheses*

Consider the following statistical model:

$$\{p(x;\theta); \theta \in \Theta = \{\theta_0, \theta_1\}\} \qquad [2.17]$$

characterized by a set Θ that only consists of two values $\theta_0 \neq \theta_1$ and the hypothesis $H_0 = \{\theta_0\}$. We let:

$$\Lambda(x) = \frac{p(x;\theta_1)}{p(x;\theta_0)} \qquad [2.18]$$

The following result, obtained by Neyman and Pearson, is fundamental, which gives the expression of the UMP test at the level α [NEY 33].

THEOREM 2.1 (Neyman and Pearson).– For any value of α, there are two constants $\eta \geq 0$ and $f \in (0,1)$, such that the critical function test

$$T^*(X) = \begin{cases} 1 & \text{if } \Lambda(X) > \eta \\ f \in (0,1[& \text{si } \Lambda(X) = \eta \\ 0 & \text{if } \Lambda(X) < \eta \end{cases} \qquad [2.19]$$

is UMP at the level α. When $\Lambda(X) = \eta$, the value of T is given by an auxiliary Bernoulli random variable with the probability f of being equal to 0. The values of η and f can be derived from the confidence level expression:

$$\mathbb{E}_{\theta_0}\{T^*(X)\} = \alpha \qquad [2.20]$$

The function $\Lambda(X)$ is known as the likelihood ratio and the test is said to be "randomized".

EXAMPLE 2.3 (Poisson distribution test).– Consider n independent integer random variables with the Poisson distribution, which can be written as:

$$\mathbb{P}\{X_k = x\} = \frac{\lambda^x}{x!} e^{-\lambda}$$

where $x \in \mathbb{N}$ and $\lambda \in \{\lambda_0, \lambda_1\}$, with $\lambda_0 < \lambda_1$. Determine the UMP test of $H_0 = \{\lambda_0\}$, i.e. performing η and f given α.

HINTS: *The log-likelihood ratio is written as:*

$$\ell = -n(\lambda_1 - \lambda_0) + \log(\lambda_1/\lambda_0) \sum_{k=0}^{n-1} X_k$$

Comparing ℓ to a given threshold is equivalent to comparing $\Lambda(X) = \sum_{k=0}^{n-1} X_k$ to another threshold. Hence, the UMP test is written as:

$$T^*(X) = \begin{cases} 1 & \text{if } \Lambda(X) > \eta \\ f \in (0,1) & \text{si } \Lambda(X) = \eta \\ 0 & \text{if } \Lambda(X) < \eta \end{cases} \quad [2.21]$$

Using the characteristic function, it is easy to show that the distribution of Λ under H_0 is Poisson with parameter $\phi = n\lambda_0$.

For a given value of the significance level α, the values of η and f are derived from the equation $\alpha = \mathbb{P}_{\lambda_0}\{\Lambda > \eta\} = 1 - \mathbb{P}_{\lambda_0}\{\Lambda \leq \eta\}$, which can be written as:

$$\alpha = 1 - \sum_{x=0}^{\eta-1} \frac{\phi^x}{x!} e^{-\phi} - f \frac{\phi^\eta}{\eta!} e^{-\phi}$$

Hence:

$$\eta = \min\{x \in \mathbb{N} \text{ s.t. } 1 - \sum_{x=0}^{\eta-1} \frac{\phi^x}{x!} e^{-\phi} \leq \alpha\} \text{ and } f = \frac{\alpha - (1 - \sum_{x=0}^{\eta-1} \frac{\phi^x}{x!} e^{-\phi})}{\frac{\phi^\eta}{\eta!} e^{-\phi}}$$

∎

In a large number of practical cases, the probability of the random variable $\Lambda(X)$ being exactly equal to η is 0. The most typical case is when the distribution of Λ has a density. In this case, the critical function is written as:

$$T^*(X) = \begin{cases} 1 & \text{if } \Lambda(X) > \eta \\ 0 & \text{if } \Lambda(X) \leq \eta \end{cases} \quad [2.22]$$

This test is said to be "deterministic".

The fundamental result is that the optimal test is based on the likelihood ratio.

The inequality $\Lambda(X) \geq \eta$ can often be simplified, as in the case of exponential models. This may be illustrated using the following example. Consider the model $\{$ i.i.d. $\mathcal{N}(n; m, 1)\}$, with $m \in \{m_0, m_1\}$ and $m_1 > m_0$. The likelihood ratio is written as:

$$\Lambda(X) = \frac{p(x; \theta_1)}{p(x; \theta_0)} = \exp\left((m_1 - m_0) \sum_{k=0}^{n-1} X_k - \frac{n}{2}(m_1^2 - m_0^2)\right)$$

However, as the exponential function is monotonic, comparing $\Lambda(X)$ to a threshold is equivalent to comparing the argument to another threshold. This comes down to comparing to a threshold of the quantity:

$$\Psi(X) = 2(m_1 - m_0) \sum_{k=0}^{n-1} X_k - n(m_1^2 - m_0^2)$$

This can be simplified further by comparing the following statistic to a threshold ζ:

$$T(X) = \mathbb{1}(\Phi(X) \geq \zeta) \quad \text{with} \quad \Phi(X) = \frac{1}{n} \sum_{k=0}^{n-1} X_k \qquad [2.23]$$

The statistic Φ has a much simpler expression than that used for the likelihood ratio. Under the hypothesis H_0, the random variable $\Phi(X)$ is Gaussian, with mean m_0 and variance $1/n$. The threshold value is then determined in such a way as to satisfy the level α, which is written as:

$$\zeta = m_0 + Q^{[-1]}(\alpha)\sqrt{n}$$

where $Q^{[-1]}(\alpha)$ is the inverse cumulative function of the standard Gaussian distribution.

Note that the UMP test at the level α is not dependent on m_1.

Figure 2.5 shows the connection between the significance level (or probability of false alarm), the power (or probability of detection) and the probability densities of the random variable $\Phi(X)$ under the two hypotheses.

Obviously, increasing the threshold value reduces the probability of a false alarm, but also reduces the probability of detection. A compromise between the two types of error is therefore required (see example 2.4).

For some issues, a cost is assigned to each decision. A cost function is defined as an application of $\Theta \times \Theta \mapsto \mathbb{R}^+$. In the very simple case where Θ consists of two values, the cost function takes four values $C(\theta_i, \theta_j) \in \mathbb{R}^+$ with i, j in $\{0, 1\}$. The Bayesian approach consists of determining the threshold that minimizes the mean risk:

$$R_B = \sum_{i=0}^{1} \sum_{j=0}^{1} C(\theta_i, \theta_j) \mathbb{P}\{\text{decide } i \mid \text{knowing it is } j\} \qquad [2.24]$$

The choice of the cost function depends on the objective of the application, and should be made with assistance from the experts of the domain of interest. A common non-informative choice is $C(\theta_i, \theta_j) = 1 - \delta_{i,j}$.

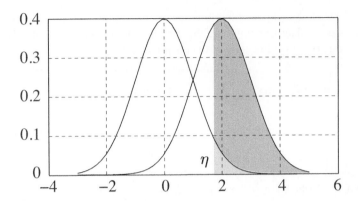

Figure 2.5. *The two curves represent the respective probability densities of the test function $\Phi(X)$ under the hypothesis H_0 ($m_0 = 0$) and hypothesis H_1 ($m_1 = 2$). For a threshold value η, the light gray area represents the significance level or probability of false alarm, and the dark gray area represents the power or probability of detection*

EXAMPLE 2.4 (ROC curve).– Let $\{$ i.i.d. $\mathcal{N}(n; m, 1)\}$ be a statistical model, where $m \in \{m_0, m_1\}$ with $m_1 > m_0$. We will test the hypothesis $H_0 = \{m_0\}$. The UMP test at the level α compares:

$$\Phi(X) = \frac{1}{n} \sum_{k=0}^{n-1} X_k$$

with a threshold ζ. Let us note that the variance of $\Phi(X)$ is equal to $1/n$ for any values of m. The value of ζ is derived from the value α of the significance level. Then, using this value of ζ, we perform the power β of the test. The curve giving β as a function of α is known as the receiving operational characteristic (ROC) curve. In our case, its expression as a function of the parameter $\zeta \in \mathbb{R}$ is written as:

$$\begin{cases} \alpha = \int_{\zeta}^{+\infty} \frac{1}{\sigma\sqrt{2\pi}} e^{-(t-m_0)^2/2\sigma^2} dt \\ \beta = \int_{\zeta}^{+\infty} \frac{1}{\sigma\sqrt{2\pi}} e^{-(t-m_1)^2/2\sigma^2} dt \end{cases} \quad [2.25]$$

where $\sigma^2 = 1/n$. Figure 2.6 shows the ROC curve for $m_0 = 0$ and $m_1 = 1$ and different values of n. The form of these curves is typical. The ROC curve is increasing and is concave above the first bisector. Note that the first bisector is the ROC curve associated with the purely random test, which accepts the hypothesis H_1 with probability α. Hence, for a given significance level α, the power β may not be under the level α. The closer the ROC curve to the point $(0, 1)$, the more efficient the detector in discriminating between the two hypotheses. One way of characterizing this efficiency is to calculate the area under the ROC curve, known as the AUC (area under the ROC curve). The expression of the AUC associated with the test function S is given by:

$$A = \int_{-\infty}^{+\infty} \beta(\zeta) d\alpha(\zeta)$$
$$= \int_{-\infty}^{+\infty} \int_{-\infty}^{+\infty} \mathbb{1}(u_0 < u_1) p_S(u_0; \theta_0) p_S(u_1; \theta_1) du_0 du_1 \qquad [2.26]$$

where $p_S(u_0; \theta_0)$ and $p_S(u_1; \theta_1)$ are the respective distributions of S under the two hypotheses.

Expression [2.26] can be interpreted as the probability that U_0 is lower than U_1 for two independent random variables, U_0 and U_1, with respective distributions, $p_S(u_0; \theta_0)$ and $p_S(u_1; \theta_1)$.

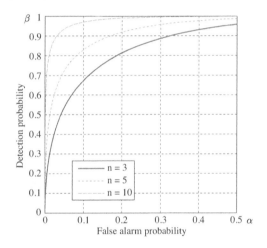

Figure 2.6. *ROC curve. The statistical model is { i.i.d. $\mathcal{N}(n; m, 1)$}, where $m \in \{0, 1\}$. The hypothesis $H_0 = \{0\}$ is tested by the likelihood ratio [2.18]. The higher the value of n, the closer the curve to the ideal point, with coordinates $(0, 1)$. The significance level α is interpreted as the probability of a false alarm. The power β is interpreted as the probability of detection*

2.5.1.1. *Experimental ROC curve and AUC*

The ROC curve and the AUC are valuable tools for evaluating test functions. However, it is important to note that in the case of composite hypotheses, the power depends on the choice of the parameter in the hypothesis H_1. There is therefore an infinite number of possible ROC curves, each associated with a value of parameter $\theta \in H_1$. We may therefore choose a sub-set of values of H_1 beforehand, carrying out a random draw in order to obtain a mean curve.

An experimental approach may also be used. To do this, a series of measurements is carried out, including N_0 examples under the hypothesis H_0 and N_1 examples under the hypothesis H_1. These data are then used to evaluate the test statistic Φ in both cases. We obtain a series of values $\Phi_{n_0|0}$, with n_0 from 1 to N_0 for the data labeled H_0, and $\Phi_{n_1|1}$, with n_1 from 1 to N_1 for the data labeled H_1. Using these two series of values, it is possible to estimate the experimental ROC curve via the false alarm probability estimator $\hat{\alpha} = N_0^{-1} \sum_{n_0=0}^{N_0-1} \mathbb{1}(\Phi_{n_0|0} > \eta)$ and the detection probability estimator $\hat{\beta} = N_1^{-1} \sum_{n_1=0}^{N_1-1} \mathbb{1}(\Phi_{n_1|1} > \eta)$ (see exercise 2.2). To obtain an estimation $\hat{A}$ of the area under the ROC curve, expression [2.26] can be used to derive the Mann-Whitney statistic, which is written as:

$$\hat{A} = \frac{1}{N_0 N_1} \sum_{n_0=0}^{N_0-1} \sum_{n_1=0}^{N_1-1} \mathbb{1}(\Phi_{n_1|1} \geq \Phi_{n_0|0}) \qquad [2.27]$$

Note that the choice of the database, containing examples under both H_0 and H_1, is critical: this database must include a sufficient number of examples that are representative of the application in question.

EXERCISE 2.2 (Empirical ROC curve and AUC).– (see p. 172) Consider the statistical model $\{$ i.i.d. $\mathcal{N}(n; m, 1)\}$, where $m \in \{m_0, m_1\}$, with $m_1 > m_0$, and the hypothesis $H_0 = \{m_0\}$. An observation consists of n values. We consider that a database is available containing N_0 observations under H_0 and N_1 observations under H_1. In this exercise, these observations will be used to estimate the ROC curve and the AUC associated with the statistic [2.23], even though analytical expressions [2.25] are available in this specific case:

1) Using the expression [2.23], propose an estimator of the significance level and an estimator of the power, which are based on the database.

2) Write a program to:

1) draw $N_0 = 1000$ observations of length $n = 10$ for H_0 and $N_1 = 1200$ observations of length $n = 10$ for H_1,

2) estimate the ROC curve and compare it to the theoretical expression given by [2.25],

3) estimate the area under the ROC curve with the Mann-Whitney statistic.

2.5.2. *Generalized likelihood ratio test (GLRT)*

Let us consider the parametric statistical model $\{P_\theta; \theta \in \Theta \subset \mathbb{R}^p\}$. In the following, we shall assume that this family is dominated, and hence P_θ has a probability density which may be simply denoted as $p(x; \theta)$. The basic hypothesis H_0 and the alternative H_1 are both assumed to be composite.

In the general case, the UMP test does not exist. For example, consider a situation where the parametric family is dependent on a single scalar parameter θ and where the hypothesis under test is $H_0 = \{\theta \leq \theta_0\}$. If the UMP test at the level α of the simple hypothesis $\{\theta_0\}$ against the simple hypothesis $\{\theta_1\}$, with $\theta_1 > \theta_0$, does depend on θ_1, then a UMP test cannot be carried out.

In practice, in the absence of a UMP test, the generalized likelihood ratio test (GLRT) is used, although it has attracted a good deal of criticism.

The GLRT is based on the critical function $T(X) = \mathbb{1}(\Lambda(X) > \eta)$, with the following test statistic:

$$\Lambda(X) = \frac{\max_{\theta \in \Theta} p(X;\theta)}{\max_{\theta \in H_0} p(X;\theta)} \quad [2.28]$$

This situation is illustrated in Figure 2.7. The larger the value of $\Lambda(X)$ compared to 1, the greater the probability of rejecting the hypothesis H_0.

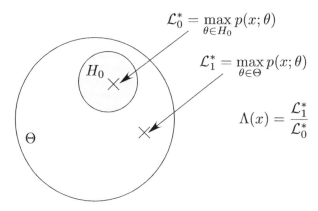

Figure 2.7. *Diagram showing the calculation of the GLRT*

Note that $\Lambda(x)$ is positive. We may therefore take the logarithm and define as:

$$\mathcal{L}(X) = 2 \log \Lambda(X) = 2 \left(\max_{\theta \in \Theta} \log p(X;\theta) - \max_{\theta \in H_0} \log p(X;\theta) \right) \quad [2.29]$$

Subsequently, as the logarithmic function is monotonous and increasing, comparing $\Lambda(X)$ to the threshold η is equivalent to comparing $\mathcal{L}(X)$ to the threshold $\log \eta$.

Taking $\theta = (\theta_0, \ldots, \theta_{q-1}, \theta_q, \ldots, \theta_{p-1})$ and the hypothesis under test:

$$H_0 = \{\theta : \theta_0 = \ldots = \theta_{q-1} = 0, \theta_q, \ldots, \theta_{p-1}\}$$

under relatively general conditions [MON 10], the test function [2.29] for H_0 may be shown to asymptotically follow a χ^2 distribution with q degrees of freedom, which is written as:

$$\mathcal{L}_{H_0}(X) \sim \chi_q^2 \qquad [2.30]$$

This result is used in exercises 2.8 and 2.32.

In the specific case where $q = 1$, the hypothesis H_0 to be tested is often:

– of the form $H_0 = \{\theta_1 \leq \mu_0\}$, where μ_0 is a given value. The hypothesis is said to be unilateral. It is often tested using a unilateral test;

– of the form $H_0 = \{\mu_0 \leq \theta_1 \leq \mu_1\}$, where μ_0 and μ_1 are given values. In this case, the hypothesis is said to be bilateral. One specific case occurs when $\mu_0 = \mu_1$, which gives the hypothesis $H_0 = \{\mu = \mu_0\}$. Bilateral hypothesis testing is often carried out in these cases. This case is illustrated by the test presented in the following section.

2.5.2.1. *Bilateral mean testing*

Consider the statistical model $\{$ i.i.d. $\mathcal{N}(n; m, \sigma^2)\}$, where $(m, \sigma^2) \in \Theta = \mathbb{R} \times \mathbb{R}^+$ and with a hypothesis $H_0 = \{m_0\} \times \mathbb{R}^+$. The GLRT at the significance level α is written as:

$$\frac{\sqrt{n}|m_0 - \widehat{m}|}{(n-1)^{-1/2}\sqrt{\sum_{k=0}^{n-1}(X_k - \widehat{m})^2}} \underset{H_0}{\overset{H_1}{\gtrless}} T_{n-1}^{[-1]}(1 - \alpha/2) \qquad [2.31]$$

where $\widehat{m} = n^{-1}\sum_{k=0}^{n-1} X_k$ and T_{n-1} denotes the Student distribution with $n - 1$ degrees of freedom.

PROOF.– Indeed, the log-likelihood is written as:

$$\mathcal{L}(\theta) = \log p(X; \theta) = -\frac{n}{2}\log(2\pi) - \frac{n}{2}\log(\sigma^2) - \frac{1}{2\sigma^2}\sum_{k=0}^{n-1}(X_k - m)^2$$

Canceling the first derivative with respect to σ^2, we obtain the expression of the maximum:

$$\widetilde{\mathcal{L}}(m) = -\frac{n}{2}\log(2\pi) - \frac{n}{2}\log\sum_{k=0}^{n-1}(X_k - m)^2$$

For $m = m_0$ (hypothesis H_0), the maximum is expressed as:

$$-\frac{n}{2}\log(2\pi) - \frac{n}{2}\log\sum_{k=0}^{n-1}(X_k - m_0)^2$$

and for $m \neq m_0$ (hypothesis H_1), the maximum is obtained by canceling the first derivative of $\widetilde{\mathcal{L}}(m)$ with respect to m, and is expressed as:

$$-\frac{n}{2}\log(2\pi) - \frac{n}{2}\log\sum_{k=0}^{n-1}(X_k - \widehat{m})^2$$

Consequently, following [2.28]:

$$\Lambda^{2/n}(X) = \frac{\sum_{k=0}^{n-1}(X_k - \widehat{m})^2}{\sum_{k=0}^{n-1}(X_k - m_0)^2} = \frac{\sum_{k=0}^{n-1}(X_k - \widehat{m})^2}{\sum_{k=0}^{n-1}(X_k - \widehat{m} + (m_0 - \widehat{m}))^2}$$

$$= \frac{\sum_{k=0}^{n-1}(X_k - \widehat{m})^2}{\sum_{k=0}^{n-1}(X_k - \widehat{m})^2 + n(\widehat{m} - m_0)^2}$$

$$= \frac{1}{1 + n(\widehat{m} - m_0)^2 / \sum_{k=0}^{n-1}(X_k - \widehat{m})^2} \quad [2.32]$$

Due to the monotonic nature of the function $1/(1 + u^2)$, comparing the statistic $\Lambda(X)$ to another threshold is equivalent to comparing the statistic $|V(X)|$ to a threshold, where:

$$V(X) = \frac{\sqrt{n}(\widehat{m} - m_0)}{\sqrt{(n-1)^{-1}\sum_{k=0}^{n-1}(X_k - \widehat{m})^2}} \quad [2.33]$$

Setting $p = 0$ and then $\beta_0 = m_0$ in equation [2.83], we shall establish that, under H_0, $V(X)$ follows a Student distribution with $(n-1)$ degrees of freedom. The test of H_0 at the level α is therefore written as [2.31]. ∎

EXERCISE 2.3 (Student distribution for H_0).– (see p. 166) Taking the statistical model { i.i.d. $\mathcal{N}(n; m, \sigma^2)$} and the hypothesis $H_0 = \{m_0\}$, write a program to:

– simulate, for H_0, 10000 observations of length $n = 100$;

– calculate $V(X)$ following expression [2.33];

– compare the histogram of $V(X)$ to the theoretical curve given by Student's distribution. The t.pdf function in *Python®* provides the probability density values of Student's distribution.

EXERCISE 2.4 (Unilateral mean testing).– (see p. 167) Taking the statistical model $\{\text{i.i.d.}\ \mathcal{N}(n; m, \sigma^2)\}$, where $(m, \sigma^2) \in \Theta = \mathbb{R} \times \mathbb{R}^+$ and the hypothesis $H_0 = \{m \geq m_0\} \times \mathbb{R}^+$, we determine the GLRT of H_0 at the significance level α.

2.5.2.2. Test p-value

Let us consider a test of the hypothesis H_0, for which the critical function is of monolateral form, $T(X) = \mathbb{1}(S(X) > \eta)$, where $S(X)$ is a real-valued function. Based on the observation X, it is possible to calculate $T(X)$, which takes a value of either 0 or 1. We will determine a confidence level associated with this decision. The first step is to calculate the value of $S(X)$; however, this value is meaningless in absolute terms, whereas the probability of observing a value greater than $S(X)$ under the hypothesis H_0 has a clear meaning: the lower the probability, the more reasonable it would be to reject H_0. The probability distribution of $S(X)$ for all values of the parameter $\theta \in H_0$ is necessary in order to carry out this calculation. If we denote $p_S(s; \theta)$ as the probability density of this distribution, the following statistic is known as the p-value:

$$p\text{-value} = \min_{\theta \in H_0} \int_{S(X)}^{+\infty} p_S(s; \theta) ds \qquad [2.34]$$

The closer this value to 0, the more reasonable it would be to reject H_0.

Clearly, if the critical function is of bilateral form, $T(X) = \mathbb{1}(S(X) \notin (-\eta, +\eta))$, the p-value is defined by:

$$p\text{-value} = 2 \min_{\theta \in H_0} \int_{|S(X)|}^{+\infty} p_S(s; \theta) ds \qquad [2.35]$$

Note that a common error consists of comparing the p-values of two samples of different sizes. This is irrelevant because, generally, as the size of a sample increases, the distribution of $S(X)$ narrows and thus the p-value decreases. Moreover, note that the p-value is dependent on the chosen test statistic. Consequently, for the same observations, it may take a high value for one test statistic and a low value for another statistic. Its meaning is therefore often debated, but this technique remains widespread. Typically, rejection of H_0 is recommended in cases with a p-value of less than 1%.

2.5.2.3. Confidence interval associated with a test

DEFINITION 2.7.– Consider a statistical model $\{P_\theta; \theta \in \Theta\}$ defined over $\mathcal{X}$ and a sequence of observations $X = \{X_0, \ldots, X_{n-1}\}$. Let $\alpha \in (0, 1)$. The confidence region for θ at $100(1-\alpha)\%$ is a region $\mathcal{I}_\alpha(X) \subset \Theta$, such that, for any θ:

$$\mathbb{P}_\theta \{\theta \in \mathcal{I}_\alpha(X)\} \geq 1 - \alpha$$

When θ is of dimension 1, we speak of a confidence interval.

A test may therefore be associated with a region of confidence, which is as follows. Taking a region of confidence $\mathcal{I}_\alpha(X)$, at $100(1-\alpha)\%$ of θ, the indicator function $\mathbb{1}(\theta_0 \in \mathcal{I}_\alpha(X))$ appears as a test function of the hypothesis $H_0 = \{\theta = \theta_0\}$ at the significance level α.

Reciprocally, consider a statistical model, and let $R_\alpha(\theta_0) \subset \mathcal{X}$ be the critical region of a test of significance level α associated with the hypothesis $H_0 : \{\theta = \theta_0\}$. In this case, each value $x \in \mathcal{X}$ may be associated with a set $\mathcal{S}_\alpha(x) \subset \Theta$ defined by:

$$\mathcal{S}_\alpha(x) = \{\theta_0 \in \Theta \quad \text{s.t.} \quad R_\alpha(\theta_0) \ni x\}$$

As the test is of level α, $\mathbb{P}_{\theta_0}\{X \notin R_\alpha(\theta_0)\} \leq \alpha$ and thus:

$$1 - \alpha \leq \mathbb{P}_{\theta_0}\{X \in R_\alpha(\theta_0)\} = \mathbb{P}_{\theta_0}\{\mathcal{S}_\alpha(X) \ni \theta_0\}$$

meaning that $\mathcal{S}_\alpha(X)$ is a region of confidence at $100(1-\alpha)\%$ of θ_0.

EXAMPLE 2.5 (Variance confidence interval).– We consider the statistical model $\{\text{ i.i.d. } \mathcal{N}(n; m, \sigma^2)\}$, with unknown mean and variance, and determine the GLRT at the significance level α associated with the hypothesis $H_0 = \{\sigma = \sigma_0\}$. We then use this result to deduce a confidence interval at $100(1-\alpha)\%$ of σ^2.

HINTS: *Based on* [2.28], *after simplification, we obtain the following test statistic:*

$$\Phi(X) = \frac{S(X)}{\sigma_0^2}$$

where $S(X) = \sum_{k=0}^{n-1}(X_k - \widehat{m})^2$, *with* $\widehat{m} = n^{-1}\sum_{k=0}^{n-1} X_k$. *Using the results of property 2.3*, $\Phi(X)$ *may be shown to follow a* χ_{n-1}^2 *distribution with* $(n-1)$ *degrees of freedom.*

We may then identify a critical region at the level α:

$$R_\alpha(\sigma_0) = \{X \in \mathbb{R}^n \quad \text{s.t.} \quad \sigma_0^2 \chi_{n-1}^{2,[-1]}(\alpha/2) \leq S(X) \leq \sigma_0^2 \chi_{n-1}^{2,[-1]}(1 - \alpha/2)\}$$

where $\chi_{n-1}^{2,[-1]}$ *is the inverse of the cumulative function of the* χ^2 *distribution with* $(n-1)$ *degrees of freedom. From this, we can deduce a confidence interval at* $100(1-\alpha)\%$ *of* σ^2:

$$\mathcal{I}_\alpha(X) = \left\{\sigma^2 \in \mathbb{R}^+ \quad \text{s.t.} \quad \frac{S(X)}{\chi_{n-1}^{2,[-1]}(1-\alpha/2)} \leq \sigma^2 \leq \frac{S(X)}{\chi_{n-1}^{2,[-1]}(\alpha/2)}\right\}$$

∎

EXERCISE 2.5 (Mean equality test).– (see p. 168) Consider two samples $\{X_{0,0}, \ldots, X_{0,n_0-1}\}$ and $\{X_{1,0}, \ldots, X_{1,n_1-1}\}$ i.i.d. Gaussian, independent, with respective means m_0 and m_1, a common variance σ^2, and respective lengths n_0 and n_1. Here σ is presumed to be unknown. We determine a test for the hypothesis $m_0 = m_1$.

1) Describe the statistical model and the hypothesis H_0;

2) Determine the GLRT of H_0 at the significance level α;

3) Show that, for H_0, the test statistic is expressed as the modulus of a Student r.v. with $(n_0 + n_1 - 2)$ degrees of freedom. Use this result to identify the test at the level α;

4) Use this result to determine the p-value;

5) Consider Table 2.1, which provides the growth results of a plant (in mm) using two different fertilizers, denoted as 1 and 2. Give the p-value of the mean equality test and present your conclusions.

F	1	1	1	1	2	2	2
size (mm)	51.0	53.3	55.6	51.0	55.5	53.0	52.1

Table 2.1. *Growth in mm under the action of 2 fertilizers*

EXERCISE 2.6 (CUSUM algorithm).– (see p. 170) Consider a sequence of n independent random variables. These variables are presumed to have the same probability distribution, of density $p(x; \mu_0)$, between instants 0 and $m - 1$ and the same probability distribution, of density $p(x; \mu_1)$, between instants m and $n - 1$, where m may take any value from 0 to $n - 2$ inclusive, or keep the same distribution $p(x; \mu_0)$ over the whole sample. The value of m is unknown, but both μ_0 and μ_1 are known.

We will test the hypothesis H_0 that no sudden jump exists from μ_0 toward μ_1:

1) Describe the statistical model;

2) Determine the test function T_n of the GLRT of the hypothesis H_0, associated with a sample of size n;

3) Taking $s_k = \log p(x_k; \mu_1)/p(x_k; \mu_0)$, $C_n = \sum_{k=0}^{n-1} s_k$, show that T_n may be calculated recursively using the form $T_k = \max\{0, T_{k-1} + s_k\}$. Write a program to verify this assertion;

4) Write a program to implement the CUSUM algorithm.

As this test is carried out using the cumulated sum of likelihood ratios, it is generally referred to as the CUSUM. When the parameter μ_0 is unknown, it can be estimated using a recursive mean algorithm. When the parameter μ_1 is unknown, the situation is clearly more critical. Reference [BAS 93] gives a full treatment of the field.

2.5.3. χ^2 goodness-of-fit test

A *goodness-of-fit test* is a test that is concerned with the basic hypothesis H_0 that the observed sample comes from a given probability distribution. As mentioned previously, the hypothesis $H_0 = \{P = P_0\}$ may be simple or composite. Hence, if the hypothesis H_0 is that the observation follows a Gaussian distribution with a given mean and variance, then H_0 is simple. However, if the variance is unknown, then H_0 is composite. Evidently, if the set of possible probability distributions under consideration cannot be indexed using a parameter of finite dimensions, then the model will be non-parametric.

Consider a sequence of N real random variables, $\{X_0, \ldots, X_{N-1}\}$, i.i.d. and a partition of $\mathbb{R}$ into g disjoint intervals $\Delta_0, \ldots, \Delta_{g-1}$. Let us denote:

$$N_j = \sum_{k=0}^{N-1} \mathbb{1}(X_k \in \Delta_j) \qquad [2.36]$$

which represent the number of values in the sample $\{X_1, \ldots, X_{N-1}\}$, which are located within the interval Δ_j. Note that $N_0 + \cdots + N_{g-1} = N$, and $p_{j,0} = \mathbb{P}_{H_0}\{X_k \in \Delta_j\}$ with $\sum_j p_{j,0} = 1$. In cases where the hypothesis H_0 is composite, the quantities $p_{j,0}$ may be dependent on unknown parameters. In this case, these parameters are replaced by consistent estimations. In exercise 2.7, for example, the true value of the variance is replaced by an estimation.

Exercise 2.7 shows that the random variable:

$$X^2 = \sum_{j=0}^{g-1} \frac{(N_j - np_{j,0})^2}{np_{j,0}} = N \sum_{j=0}^{g-1} \frac{\left(\frac{N_j}{N} - p_{j,0}\right)^2}{p_{j,0}} \xrightarrow{d} \chi^2_{g-1} \qquad [2.37]$$

This result has an obvious meaning: X^2 measures the relative distance between the empirical frequencies N_j/N and the theoretical probabilities. Based on the law of large numbers 1.8, N_j/N converges under H_0 toward $p_{j,0}$ and under H_1 toward $p_{j,1} \neq p_{j,0}$. Thus, the statistic X^2 will take small values for H_0 and large values for H_1; these values will be increasingly large as N increases. This justifies the use of an unilateral test, which compares X^2 to a threshold.

The choice of the number g and the interval sizes Δ_j is critical. One method consists of selecting g classes of the same probability, or dividing the set of sample values into classes with the same empirical weight (see exercise 2.8).

EXERCISE 2.7 (Proof of [2.37]).– (see p. 172)

Using the conditions associated with expressions [2.36] and [2.37], take $D = \text{diag}(P)$ with $P = [p_0]^T$, where $p_j = \mathbb{P}\{X_k \in \Delta_j\}$:

1) Determine the expressions of $\mathbb{E}\{N_j\}$ and $\mathbb{E}\{N_j N_m\}$ as a function of p_j. Use these expressions to deduce the fact that the covariance matrix of the random vector $(N_0, \ldots, N_{g-1})$ can be written as nC, where C is a matrix for which the expression will be given.

2) Show that $D^{-1/2}CD^{-1/2}$ is a projector of rank $(g-1)$.

3) Taking $\widehat{P}_j = N_j/N$ and $Y_j = \sqrt{N}(P_j - p_j)$, determine the asymptotic distribution of the random vector $Y = [Y_0 \ldots Y_{g-1}]^T$. Use this result to determine the asymptotic distribution of vector $Z = D^{-1/2}Y$.

Finally, identify the asymptotic distribution of variable $X^2 = \sum_{j=0}^{g-1} Z_j^2$.

EXERCISE 2.8 (Chi2 fitting test).– (see p. 172) Write a program to simulate the hypothesis $H_0 = \{m = 0\}$ for the model $\{$ i.i.d. $\mathcal{N}(N; m, \sigma^2)\}$. To do this, 3000 drawings of a sample of size $N = 200$ are required, with a Gaussian random variable of mean 0 and unknown variance σ^2. The set of values of each draw should be separated into $g = 8$ blocks containing the same number of samples, i.e. $200/8$. Using the cumulative function of a Gaussian, calculate the values of $p_j = \mathbb{P}\{X_k \in \Delta_j\}$, in which the true, unknown value of σ is replaced by the estimation given by the numpy.std function. Using this result, the values of X^2 can be deduced, given by expression [2.37], and a histogram is constructed, which will be compared to the theoretical curve of χ^2 with $g - 1 = 7$ degrees of freedom.

2.6. Statistical estimation

2.6.1. *General principles*

In this section, we will consider the problem of *parameter estimation*. We will begin by presenting a number of general results concerning performance evaluation, notably bias and quadratic error. We will then examine three methods used in constructing estimators: the least squares method, mainly in the case of the linear model, the moment method, and the maximum likelihood approach.

DEFINITION 2.8 (Estimator).– Let $\{P_\theta; \theta \in \Theta\}$ be a statistical model of the observation X. An estimator is any (measurable) function of X with values in Θ.

One fundamental question concerns the evaluation of estimator quality. We try to choose an estimator $\widehat{\theta}$, such that $\mathbb{P}_\theta\left\{\widehat{\theta} \neq \theta\right\} = 0$. Estimators of this type only exist in exceptional situations and have no practical interest. In practice, two quantities are commonly used, the bias and the risk, as defined below.

DEFINITION 2.9 (Bias and risk of an estimator).– Consider an estimator $\widehat{\theta} : \mathcal{X} \mapsto \Theta \subset \mathbb{R}^d$. The bias is the vector of dimension d:

$$B(\theta,\widehat{\theta}) = \mathbb{E}_\theta\left\{\widehat{\theta}\right\} - \theta$$

The quadratic risk is the square matrix of dimension d:

$$R(\theta,\widehat{\theta}) = \mathbb{E}_\theta\left\{(\theta - \widehat{\theta})(\theta - \widehat{\theta})^T\right\} \qquad [2.38]$$

It is easy to show that

$$R(\theta,\widehat{\theta}) = \operatorname{cov}_\theta\left\{\widehat{\theta}\right\} + B(\theta,\widehat{\theta})B^T(\theta,\widehat{\theta})$$

where

$$\operatorname{cov}_\theta\left\{\widehat{\theta}\right\} = \mathbb{E}_\theta\left\{\left(\widehat{\theta} - \mathbb{E}\left\{\widehat{\theta}\right\}\right)\left(\widehat{\theta} - \mathbb{E}\left\{\widehat{\theta}\right\}\right)^T\right\}$$

is the covariance matrix of $\widehat{\theta}$.

It is worth noting that an estimator does not depend on the unknown parameter being estimated, but the performance of the estimator does depend on this parameter. Thus, the bias and the risk are generally dependent on the unknown value θ.

It is pointless to try and find an estimator with the minimum quadratic risk for all values of θ. For this reason, we have to restrict the search class for $\widehat{\theta}$. For example, we can limit the search to bias-free estimators, or to the class of linear estimators, or the class of estimators that are invariant by translation, etc.

Another method involves a Bayesian approach, which consists of taking into account the available knowledge concerning θ, which takes the form of a probability distribution $p_\Theta(\theta)$ and may be used for minimizing the average risk:

$$R_B(\widehat{\theta}) = \int_\Theta \operatorname{trace}\left\{R(\theta,\widehat{\theta})\right\} p_\Theta(\theta) d\theta \qquad [2.39]$$

over the set of all estimators.

2.6.1.1. *Cramer-Rao bound*

Quadratic risk, expression [2.38], has a fundamental lower-bound known as the Cramer-Rao bound (CRB).

THEOREM 2.2 (Cramer-Rao bound (CRB)).– Any estimator $\widehat{\theta}$ of the parameter $\theta \in \Theta \subset \mathbb{R}^d$ verifies[2]:

$$R(\theta, \widehat{\theta}) \geq (I_d + \partial_\theta B(\theta, \widehat{\theta})) F^{-1}(\theta)(I_d + \partial_\theta B(\theta, \widehat{\theta}))^T \\ + B(\theta, \widehat{\theta}) B^T(\theta, \widehat{\theta}) \qquad [2.40]$$

where $\partial_\theta B(\theta, \widehat{\theta})$ is the Jacobian matrix of the vector $B(\theta, \widehat{\theta})$ with respect to θ, and where

$$F(\theta) = -\mathbb{E}_\theta \left\{ \partial_\theta^2 \log p(X; \theta) \right\} \qquad [2.41]$$

where $\partial_\theta^2 \log p(X; \theta)$ is the Hessian of $\log p(X; \theta)$ (square matrix of dimension d) for which the element in line m, column ℓ is written as $\frac{\partial^2 \log p(X;\theta)}{\partial \theta_m \partial \theta_\ell}$.

It can be shown that:

$$F(\theta) = \mathbb{E}_\theta \left\{ \partial_\theta \log p(X; \theta) \partial_\theta^T \log p(X; \theta) \right\} \qquad [2.42]$$

where $\partial_\theta \log p(X; \theta)$ is the Jacobian of $\log p(X; \theta)$ (vector of length d).

In the case of an unbiased estimator, formula [2.40] is written as:

$$R(\theta, \widehat{\theta}) = \mathrm{cov}_\theta \left\{ \widehat{\theta} \right\} \geq F^{-1}(\theta) \qquad [2.43]$$

The matrix F is known as the *Fisher information matrix* (FIM). An estimator that reaches the CRB is said to be *efficient*. Unfortunately, an efficient estimator does not always exist.

In the case where X is a sequence of N i.i.d. r.v. with the same density $p_X(x; \theta)$, the log-likelihood is written as $\ell(\theta) = \sum_{n=0}^{N-1} \log p_X(x_n; \theta)$ and the FIM as

$$F(\theta) = -N\mathbb{E}_\theta \left\{ \partial_\theta^2 \log p_X(X_0; \theta) \right\} \qquad [2.44]$$

The following result is given without proof.

---

[2] Consider two positive square matrices A and B of the same dimension d. We say that $A \geq B$ if and only if $A - B \geq 0$, i.e. for any $w \in \mathbb{C}^d$, we have $w^H A w \geq w^H B w$.

THEOREM 2.3.– Let us consider N observations of the Gaussian model $\{$ i.i.d. $\mathcal{N}(N; m(\theta), C(\theta))\}$, where θ is the parameter of interest. The FIM associated with this model can be written as:

$$F_{k,\ell} = \frac{N}{2}\text{trace}\left\{C^{-1}(\theta)\partial_k C(\theta)C^{-1}(\theta)\partial_\ell C(\theta)\right\} + N\partial_k m^T(\theta)C^{-1}(\theta)\partial_\ell m(\theta) \quad [2.45]$$

where $\partial_k m(\theta)$ and $\partial_k C(\theta)$, respectively, denote the derivative of m and C with respect to the kth component of θ.

It is worth noting that if $m(\theta_1)$ and $C(\theta_2)$ depend on disjoint parameters θ_1 and θ_2 of respective lengths p_1 and p_2, the FIM given by [2.45] is a two-block diagonal according to the following form:

$$F = \begin{bmatrix} F_{11} & 0_{p_1,p_2} \\ 0_{p_2,p_1} & F_{22} \end{bmatrix}$$

where F_{11} and F_{22} are square matrices, respectively, of sizes p_1 and p_2 associated with the parameters θ_1 and θ_2.

EXAMPLE 2.6 (CRB for univariate Gaussian i.i.d.).– Consider the Gaussian model $\{$ i.i.d. $\mathcal{N}(N; m, \sigma^2)\}$. Determine the CRB for the parameter $\theta = (m, \sigma^2)$.

HINTS: *The parameter of interest is (m, σ^2). Therefore, $\partial m = 1$, $\partial_{\sigma^2} m = 0$, $\partial_m \sigma^2 = 0$ and $\partial_{\sigma^2}\sigma^2 = 1$. Using the expression [2.45], we have:*

$$F_{m,m} = \frac{N}{\sigma^2},\ F_{m,\sigma^2} = 0,\ F_{\sigma^2,\sigma^2} = \frac{N}{2\sigma^4}$$

The CRB is written as:

$$F^{-1} = \begin{bmatrix} \frac{\sigma^2}{N} & 0 \\ 0 & \frac{2\sigma^4}{N} \end{bmatrix}$$

We will see in example 2.8 that $\widehat{m} = N^{-1}\sum_{n=0}^{N-1} X_n$ is unbiased with asymptotic variance σ^2/N, meaning that $\widehat{m}$ is asymptotically efficient. ∎

EXERCISE 2.9 (CRB expression using symbolic calculus).– (see p. 173) Let us consider N observations of the bivariate Gaussian model $\{$ i.i.d. $\mathcal{N}(N; m, C)\}$, where $C_{00} = \sigma_1^2$, $C_{11} = \sigma_2^2$ and $C_{01} = C_{10} = \rho\sigma_1\sigma_2$. Here ρ is the parameter of interest.

Using the symbolic toolbox of *Python*® determine the CRB (which is a square matrix of size 3) for $N = 1$. To import this toolbox, execute import sympy as sp. It is advised to use sp.Matrix, sp.diff, sp.simplify and sp.Inverse. Using the program, verify that the CRB coefficients, associated with the parameters m_1, m_2, σ_1, σ_2 and ρ, write for N i.i.d. observations:

$$\text{CRB}_{m_1} = \frac{\sigma_1^2}{N}, \quad \text{CRB}_{m_2} = \frac{\sigma_2^2}{N} \quad [2.46]$$

$$\text{CRB}_{\sigma_1} = \frac{\sigma_1^2}{2N}, \quad \text{CRB}_{\sigma_2} = \frac{\sigma_2^2}{2N}, \quad \text{CRB}_{\rho} = \frac{(1-\rho^2)^2}{N} \quad [2.47]$$

To check these results, simulation is proposed in exercise 2.21.

2.6.2. *Least squares method*

Consider a series of observations y_n of the form:

$$y_n = x_n(\beta) + w_n \quad [2.48]$$

where $x_n(\beta)$ is a deterministic model dependent on the parameter of interest $\beta \in \Theta \subset \mathbb{R}^d$ and w_n is a random process representing an additive noise. Here we deal with the context of regression analysis, where x_n is the explanatory variable and y_n is the response.

Using a sequence of N observations, the least squares method consists of the following estimator of β:

$$\widehat{\beta} = \arg\min_{\alpha \in \Theta} \sum_{n=0}^{N-1} |y_n - x_n(\alpha)|^2 \quad [2.49]$$

This very general method was introduced by Gauss to determine the trajectories of the planets. It still plays a key role in estimator construction methods. It may be applied to the condition that we have a deterministic model, dependent on the parameter being estimated. The added noise w_n models the measurement noise, but also the "model" noise, i.e. the fact that we are not completely certain of the presumed model $x_n(\beta)$. As we shall see, the least squares estimator corresponds to the maximum likelihood estimator in cases where w_n is a Gaussian random process.

Take the example of a sinusoid in noise.

EXAMPLE 2.7 (Frequency estimation of a sinusoid in noise).– Consider N observations y_n, which are written as:

$$y_n = R\cos(2\pi f_0 n - \psi) + w_n = a_c \cos(2\pi f_0 n) + a_s \sin(2\pi f_0 n) + w_n$$

where w_n is a sequence of N independent random variables with zero mean and unknown variance σ^2. The parameter is written as $\theta = (\sigma^2, f_0, a_c, a_s) \in \mathbb{R}^+ \times (0, 1/2) \times \mathbb{R} \times \mathbb{R}$. Determine the estimator of θ based on the least squares approach. We note that the model is nonlinear with respect to f_0.

Using the matrix notation and letting $a = \begin{bmatrix} a_c & a_s \end{bmatrix}^T$ and

$$H(f_0) = \begin{bmatrix} 1 & 0 \\ \cos(2\pi f_0) & \sin(2\pi f_0) \\ \vdots & \vdots \\ \cos(2\pi f_0(N-1)) & \sin(2\pi f_0(N-1)) \end{bmatrix}$$

we can write $y = H(f_0)a + w$. The least squares approach consists of minimizing $J(\theta) = \|y - H(\phi)\alpha\|^2$ with respect to θ. Then, the expression of θ is determined.

HINTS: *Canceling the derivative of J with respect to α leads to*

$$H^T(\phi)(y - H(\phi)\alpha) = 0, \quad \Rightarrow \quad \widehat{\alpha} = (H^T(\phi)H(\phi))^{-1}H^T(\phi)y$$

Carry $\widehat{\alpha}$ in J and after a few calculations, we have:

$$\widehat{f_0} = \arg \max_{\phi \in (0, 1/2)} K(\phi)$$

where $K(\phi) = y^T H(\phi)(H^T(\phi)H(\phi))^{-1}H^T(\phi)y$

This scalar maximization with respect to ϕ is highly nonlinear. However, a simple way to solve the problem consists of using a fine grid on the interval $(0, 1/2)$.

It is important to note that when the product $Nf_0 \gg 1$, the matrix $H^T(f_0)H(f_0) \approx NI_2$ and

$$K(\phi) \approx \left| \sum_{n=0}^{N-1} y_n e^{-2j\pi\phi n} \right|^2$$

is approximately equal to the squared modulus of the discrete Fourier transform (DFT). Write a program to compare the least squares estimation and the DFT estimation for different values of $f_0 N$.

Type the following program:

```
# -*- coding: utf-8 -*-
"""
Created on Sun Jun 26 18:59:40 2016
****** estimf0
@author: maurice
```

```
"""
from numpy import pi, cos, sin, arange, zeros, mean, array,
fft, sqrt, matrix as mat
from numpy.random import randn
from numpy.linalg import pinv
N = 10; sigma = 0.03; f0 = 0.08; psi = pi/3; R = 1.0;
a = array((R*cos(psi), R*sin(psi)));
L=2**12; twoL=2.0*L; err2=zeros(L); listphi = arange(L)/twoL;
Lruns = 20; hatf0 = zeros(Lruns); hatf0fft = zeros(Lruns)
for idr in range(Lruns):
    yT = R*cos(2.0*pi*f0*arange(N)-psi)+sigma * randn(N);
    y = mat(yT).T
    for ir in range(L):
        phi = listphi[ir]; H = mat(zeros([N,2]))
        H[:,0]=cos(2.0*pi*phi*arange(N).reshape(N,1))
        H[:,1]=sin(2.0*pi*phi*arange(N).reshape(N,1))
        pinvH = pinv(H.T*H); Piortho = H*pinvH*H.T
        err2[ir] = y.T*Piortho*y
    hatf0[idr] = listphi[err2.argmax()]
    Yf = abs(fft.fft(yT,2*L))**2; Yfpos = Yf[0:L]
    hatf0fft[idr] = Yfpos.argmax()/twoL
print('LS: MSE = %4.2e, FFT: MSE = %4.2e'%(sqrt(mean(abs
(hatf0-f0)**2)),\
    sqrt(mean(abs(hatf0fft-f0))**2)))
```

■

2.6.3. *Least squares method for the linear model*

Solving the problem [2.49] to find an estimator does not generally result in a simple analytical expression, but requires the use of numerical approaches. Moreover, while a number of asymptotic results exist, performance often needs to be studied on a case-by-case basis. The exception to this rule is the case of the *linear model*, for which a considerable number of results have been found. This section is devoted to a detailed study of this model, which is written as, for $n = 0, \ldots, N-1$:

$$y_n = Z_n^T \beta + \sigma \epsilon_n \qquad [2.50]$$

where $Z_n = \begin{bmatrix} 1 & X_{n,1} & \ldots & X_{n,p} \end{bmatrix}^T = \begin{bmatrix} 1 & X_n \end{bmatrix}^T$ is a sequence of known vectors, ϵ_n is a random sequence of white, zero-mean uncorrelated r.v. with variance 1 and σ is an unknown positive number.

The expression "linear model" refers to linearity of expression [2.50] with respect to the unknown parameter β. This is also known as the linear regression, where y_n denotes the explained variable or response. Explanatory or independent variables X_n may be seen in two ways: either as N points in the space $\mathbb{R}^p$ or as p points in the space $\mathbb{R}^N$.

In the following, we will consider the case of real data; however, the results are applicable only to cases involving complex data values.

2.6.3.1. *Standardization*

A common preprocessing in applied regression is the *standardization* of the explanatory variables, by subtracting its mean and dividing by its standard deviation. Subtracting the mean makes it easier to interpret the presence of interactions, similar to what we do with the covariance, and dividing by the standard deviation puts all predictors on a common scale. Let us refer to the example 2.9. For more details, see reference [GEL 08].

2.6.3.2. *Linear model assumptions*

Using the matrix notation, [2.50] can be rewritten as:

$$y = Z\beta + \sigma\epsilon, \qquad [2.51]$$

where

- y is a vector of dimension $N \times 1$;
- the so-called *design matrix*:

$$Z = \begin{bmatrix} \mathbf{1}_N & X \end{bmatrix} \qquad [2.52]$$

is a matrix of dimension $N \times (p+1)$, in which all of the components in the first column are equal to 1;

- $\beta = (\beta_0, \ldots, \beta_p) \in \mathbb{R}^{p+1}$. Coefficient β_0 is called the *intercept*;
- $\sigma \in \mathbb{R}^+$.

ϵ is a random vector taking values in $\mathbb{R}^N$ and verifying either the assumption

$$\mathbb{E}\{\epsilon\} = 0_N, \quad \mathrm{cov}(\epsilon) = I_N \qquad [2.53]$$

$$\text{or} \quad \epsilon \sim \mathcal{N}(0_N, I_N) \qquad [2.54]$$

Model [2.53] is non-parametric, whereas [2.54] is parametric.[2.54] entails [2.53], but the reverse is not necessarily true. Here the *parameter of interest* is:

$$\theta = (\beta, \sigma) \in \mathbb{R}^{p+1} \times \mathbb{R}^+ \qquad [2.55]$$

2.6.3.3. *Useful notations*

– r denotes the rank of the matrix Z. It is well known that $r \leq \min\{N, p+1\}$. If $r = (p+1) \leq N$, Z is said to be of "full column rank", and the square matrix $Z^T Z$ of dimension $p+1$ is invertible.

– Π_Z denotes the orthogonal projector onto the sub-space spanned by the columns of Z. This sub-space is denoted by $\text{Im}(Z)$ as the image of Z. Note that $0_N \leq \Pi_Z \leq I_N$ and that Π_Z has r eigenvalues equal to 1 and $(N-r)$ eigenvalues equal to 0. Hence, trace $\{\Pi_Z\} = r$.

– $\Pi_Z^\perp = I_N - \Pi_Z$ denotes the projector onto the sub-space orthogonal to $\text{Im}(Z)$. $\Pi_Z^\perp$ is positive. Hence, trace $\{\Pi_Z^\perp\} = N - r$.

– $h_{n,n}$ denotes the nth diagonal element of the matrix Π_Z. From this, we can deduce that $\sum_{n=0}^{N-1} h_{n,n} = \text{trace}\{\Pi_Z\} = r$. If u_n denotes the vector for which all components are null except for the nth component, which is 1, then $h_{n,n} = u_n^T \Pi_Z u_n$. Applying the double inequality $\Pi_{1_N} \leq \Pi_Z \leq I_N$ (see expression [2.69]) to vector u_n, we obtain:

$$\frac{1}{N} \leq h_{n,n} \leq 1 \qquad [2.56]$$

Note that $h_{n,n}$ is dependent on the X_n variables, but not on the y_n variables. The mean value of the $h_{n,n}$ is r/N. A value of $h_{n,n}$ that is close to 1 indicates that, in the space $\mathbb{R}^p$, point X_n is far from the center of the cloud of points associated with other values. Conversely, a value close to $1/N$ indicates that the point is close to this center. The quantity $h_{n,n}$ is known as the *leverage* (see exercise 2.10).

As $\Pi_Z^\perp + \Pi_Z = I_N$, any sum of the diagonal elements of the same rank in $\Pi_Z^\perp$ and Π_Z must be equal to 1, i.e. $\pi_Z^\perp(i,i) + \pi_Z(i,i) = 1$, and any sum of the non-diagonal elements of the same rank will be equal to 0, i.e. $\pi_Z^\perp(i,j) + \pi_Z(i,j) = 0$, $i \neq j$.

According to the projection theorem 2.1, the best approximation, in the least square sense, of y in the sub-space $\text{Im}(Z)$ is given by the orthogonal projection:

$$\widehat{y} = \Pi_Z y \qquad [2.57]$$

An estimator $\widehat{\beta}$ of β, in the least square sense, therefore verifies:

$$Z\widehat{\beta} = \Pi_Z y \qquad [2.58]$$

If Z is of full column rank, $Z^T Z$ is invertible and the projector

$$\Pi_Z = Z(Z^T Z)^{-1} Z^T \qquad [2.59]$$

There is therefore a single element $\widehat{\beta}$ that verifies equation [2.58], which is written as:

$$\widehat{\beta} = (Z^T Z)^{-1} Z^T y \qquad [2.60]$$

In cases where Z is not a full rank matrix, there are an infinite number of solutions defined within an additive factor, an element of the null-space (kernel) of Z.

One expression that is useful in practice can be obtained by replacing y by [2.51] in expression [2.58]:

$$Z\widehat{\beta} = Z\beta + \sigma \Pi_Z \epsilon \qquad [2.61]$$

If Z is a full column rank matrix, then [2.61] can be rewritten as:

$$\widehat{\beta} = \beta + \sigma (Z^T Z)^{-1} Z^T \epsilon \qquad [2.62]$$

This allows us to deduce the properties of $\widehat{\beta}$ from those of ϵ.

2.6.3.4. *Centering variables*

In this section, we show that the intercept β_0, on the one hand, and $\beta_1, \ldots, \beta_p$, on the other hand, are associated with two orthogonal spaces, and can therefore be calculated "separately".

Let $\mathbf{1}_N$ be a vector of length N, all the components of which are 1. We can verify that the projector of rank 1 onto the sub-space spanned by $\mathbf{1}_N$ is expressed as:

$$\Pi_{\mathbf{1}_N} = \frac{1}{N} \mathbf{1}_N \mathbf{1}_N^T$$

Let X_c be a matrix of which the column vectors are the centered column vectors of X, written as:

$$X_c = X - \Pi_{\mathbf{1}_N} X = (I_N - \Pi_{\mathbf{1}_N}) X = \Pi_{\mathbf{1}_N}^{\perp} X \qquad [2.63]$$

Finally, let Π_{X_c} be the orthogonal projector over the sub-space spanned by the columns of X_c. Multiplying expression [2.63] by $\Pi_{\mathbf{1}_N}$ on the left-hand side, we obtain $\Pi_{\mathbf{1}_N} X_c = 0$, and thus:

$$\Pi_{\mathbf{1}_N} \Pi_{X_c} = \Pi_{X_c} \Pi_{\mathbf{1}_N} = 0 \qquad [2.64]$$

This means that $\mathbf{1}_N$ is orthogonal to X_c and, consequently, $\text{Im}([\mathbf{1}_N\ X_c]) = \text{Im}(\mathbf{1}_N) \oplus \text{Im}(X_c)$. Exercise 2.10 demonstrates that:

$$\Pi_Z = \Pi_{\mathbf{1}_N} + \Pi_{X_c} \qquad [2.65]$$

From [2.65], we deduce that the orthogonal projection $\widehat{y}$, given by expression [2.57], can be written as:

$$\widehat{y} = \Pi_{\mathbf{1}_N} y + \Pi_{X_c} y \qquad [2.66]$$

Using [2.64], we have $\Pi_{\mathbf{1}_N} \widehat{y} = \Pi_{\mathbf{1}_N} y$, which can be rewritten as:

$$\frac{1}{N} \sum_{n=1}^{N} y_n = \frac{1}{N} \sum_{n=1}^{N} \widehat{y}_n \qquad [2.67]$$

Taking $y_c = y - \Pi_{\mathbf{1}_N} y$, [2.66] can also be written as:

$$\widehat{y} = \Pi_{\mathbf{1}_N} y + \Pi_{X_c} y = \Pi_{\mathbf{1}_N} y + \Pi_{X_c} y_c + \Pi_{X_c} \Pi_{\mathbf{1}_N} y = \Pi_{\mathbf{1}_N} y + \Pi_{X_c} y_c \qquad [2.68]$$

Again, this expression uses [2.64]. Expression [2.68] shows that the orthogonal projection consists of two components: one representing the mean and the other the projection onto the space spanned by the *centered* variables.

EXERCISE 2.10 (Decomposition of the design matrix).– (see p. 174)

1) Show that $(\Pi_{\mathbf{1}_N} + \Pi_{X_c})Z = Z$. Derive the expression [2.65];

2) Use this result to show that:

$$\Pi_{\mathbf{1}_N} \leq \Pi_Z \leq I_N \qquad [2.69]$$

3) Multiplying each term of [2.69] on both sides by a vector, all components of which are null except for the nth component, which is 1, demonstrate the double inequality [2.56];

4) Write a program to show that if $h_{n,n}$ is close to $1/N$, the point of coordinates X_n in $\mathbb{R}^p$ is close to the center of the cloud formed by the other points. Conversely, if $h_{n,n}$ is close to 1, the point is far from the cloud. To visualize the result with a plot, take $p = 2$.

EXERCISE 2.11 (Atmospheric CO_2 concentration).– (see p. 175) The curve shown in Figure 2.2 indicates the weekly change of the atmospheric CO_2 concentration for a period of 44 years. The data are stored in the file statsmodels.api.datasets.co2.load(). Let us note that some data are missing, annotated by NaN. We observe, on the one hand, a polynomial trend and, on the other hand, a periodic evolution. The periodic part is clearly of one year, i.e. 52 samples. We assume that the polynomial trend is of second order and the periodic part consists of the fundamental component and the second harmonic. This can be written as:

$$y_n \approx \underbrace{a_0 + a_1 n + a_2 n^2}_{\text{polynomial trend}} + \underbrace{\sum_{k=1}^{2} b_k \cos(2\pi k f_0 n) + c_k \sin(2\pi k f_0 n)}_{\text{seasonal part}} \quad [2.70]$$

with $f_0 = 7/365$. This shows that the model is linear with respect to the parameter $\theta = (a_0, a_1, a_2, b_1, b_2, c_1, c_2)$.

Write a program to determine θ and plot the residue. In theory, the residue standard deviation decreases when the order of the polynomial trend and the harmonic number increase. Therefore, it cannot be used to evaluate these two quantities. In exercise 2.39, we will see how to estimate these two values by cross-validation.

EXERCISE 2.12 (Change-point detection of Nile flow).– (see p. 175) The Nile is a major north-flowing river in north-eastern Africa. The flow between 1871 and 1970 is shown in Figure 2.8. The curve presents an apparent change point around 1899. Taking two one-order polynomial models, one before the date t and the other after the date t, determine the total error for each value of t. Determine the value of t given the minimal error.

Remark: for a sequential analysis of the change location, the CUSUM algorithm can be used (see exercise 2.6).

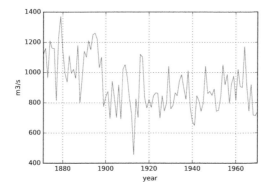

Figure 2.8. *Flow, expressed in m^3/s averaged over one year, of the Nile as measured at Ashwan from 1871 to 1970. The curve presents an apparent change around 1899*

2.6.3.5. *Probabilistic properties*

Consider the linear model given by expression [2.50] with hypotheses [2.53]. The ϵ_n are centered r.v.s, of variance 1, which are not correlated with each other. Suppose that Z is of full column rank, with the matrix $Z^T Z$ therefore being invertible.

PROPERTY 2.2 (Best Linear Unbiased Estimator - BLUE).– The estimator

$$\widehat{\beta} = (Z^T Z)^{-1} Z^T y \quad [2.71]$$

is an unbiased estimator of β and has the lowest covariance of all of the linear unbiased estimators. It is called BLUE (Best Linear Unbiased Estimator). This covariance is expressed as:

$$\mathrm{cov}\left(\widehat{\beta}\right) = \sigma^2 (Z^T Z)^{-1}$$

PROOF.– Replacing y by [2.50] in $\widehat{\beta}$, we obtain $\widehat{\beta} = \beta + \sigma(Z^T Z)^{-1} Z^T \epsilon$. If $\mathbb{E}\{\epsilon\} = 0$, then $\mathbb{E}\{\widehat{\beta}\} = \beta$. $\widehat{\beta}$ is unbiased. From this, we deduce that $\mathrm{cov}\left(\widehat{\beta}\right) = \sigma^2 (Z^T Z)^{-1}$.

Now, consider another linear estimator with respect to y of the form $\widehat{b} = Qy$, such that $\mathbb{E}\{\widehat{b}\} = \beta$. In this case, $\mathbb{E}\{\widehat{b}\} = QZ\beta = \beta$, implying that $QZ = I_{p+1}$. From this, we deduce that $\mathrm{cov}\left(\widehat{b}\right) = \sigma^2 QQ^T$.

Additionally, $A = I_N - Z(Z^T Z)^{-1} Z^T$ is a projector; $A = A^T$ and $AA = A$. Hence, $Q(I_N - Z(Z^T Z)^{-1} Z^T)Q^T \geq 0$. By developing the expression and using $QZ = I$, we deduce that:

$$QQ^T - (Z^T Z)^{-1} \geq 0$$

which is the expected result.

In the rest of this chapter, unless indicated otherwise, we shall consider hypothesis [2.54], i.e. ϵ is white and Gaussian.

PROPERTY 2.3.– The solution to problem [2.51] using hypothesis [2.54] possesses the following properties:

– $\widehat{\beta} = (Z^T Z)^{-1} Z^T y$;

– $\widehat{y} = \widehat{\beta} y \sim \mathcal{N}(Z\beta; \sigma^2 \Pi_Z)$, where the rank of Π_Z is r;

– $\widehat{e} = y - \widehat{y} \sim \mathcal{N}(0; \sigma^2 \Pi_Z^{\perp})$, where the rank of $\Pi_Z^{\perp}$ is $N - r$. We have $\mathbf{1}^T \widehat{e} = 0$;

– $\widehat{e}$ and $\widehat{y}$ are independent random variables;

– if Z is of full rank, $\widehat{e}$ and $\widehat{\beta}$ are independent random variables.

PROOF.– Substituting expression [2.51] into expression [2.57], we obtain:

$$\widehat{y} = Z\beta + \sigma \Pi_Z \epsilon \qquad [2.72]$$

using $\Pi_Z Z = Z$. From this, we deduce that the prediction residual defined by the vector $\widehat{e} = y - \widehat{y} \in \mathbb{R}^N$ can be written as:

$$\widehat{e} = y - \widehat{y} = \sigma \Pi_Z^\perp \epsilon \qquad [2.73]$$

This leads to the following results:

– based on [2.67], $\mathbf{1}^T \widehat{e} = \sum_{n=0}^{N-1} e_n = 0$;

– based on [2.72], the distribution of $\widehat{y}$ is Gaussian, which is written as:

$$\widehat{y} \sim \mathcal{N}(Z\beta; \sigma^2 \Pi_Z) \quad \Leftrightarrow \quad \frac{\widehat{y} - Z\beta}{\sigma} \sim \mathcal{N}(0; \Pi_Z) \qquad [2.74]$$

– based on [2.73], the distribution of error $\widehat{e}$ is Gaussian, which is written as:

$$\widehat{e} \sim \mathcal{N}(0; \sigma^2 \Pi_Z^\perp) \quad \Leftrightarrow \quad \frac{\widehat{e}}{\sigma} \sim \mathcal{N}(0; \Pi_Z^\perp) \qquad [2.75]$$

– the random vectors $\widehat{e}$ and $\widehat{y}$ are independent. Following [2.72] and [2.73], $\text{cov}(\widehat{e}, \widehat{y}) = \sigma^2 \Pi_Z^\perp \Pi_Z = 0$, meaning that $\widehat{e}$ and $\widehat{y}$ are not correlated; as they are jointly Gaussian, they are independent;

– if Z is of full rank, then random vectors $\widehat{e}$ and $\widehat{\beta}$ are independent. In this case, $\text{cov}(\widehat{e}, \widehat{y}) = \text{cov}(\widehat{e}, \widehat{\beta}) Z^T = 0$, hence $\text{cov}(\widehat{e}, \widehat{\beta}) = 0$. ∎

Note that the random vector $(\widehat{y} - Z\beta)$ should not be confused with the random vector $\widehat{e} = (y - \widehat{y})$. The former is not observable, as the true value of β is unknown, whereas the latter is observable (i.e. computable with the observations).

2.6.3.6. *Unbiased estimator of σ^2*

Following [2.75], $\mathbb{E}\{\widehat{e}\widehat{e}^T / \sigma^2\} = \Pi_Z^\perp$ and thus:

$$\mathbb{E}\{\widehat{e}^T \widehat{e}\} = \text{trace}\{\mathbb{E}\{\widehat{e}\widehat{e}^T\}\} = \sigma^2 \text{trace}\{\Pi_Z^\perp\} = \sigma^2 (N - r)$$

From this, we deduce that $\widehat{\sigma}^2$, defined by

$$\widehat{\sigma}^2 = \frac{1}{N-r} \widehat{e}^T \widehat{e} \qquad [2.76]$$

is an unbiased estimator of σ^2.

2.6.3.7. *Calculating a confidence interval for $Z_n\beta$*

Using expression [2.74] and multiplying the left-hand side by a vector, all components of which are null except for the nth component, which has a value of 1, we deduce that $(\widehat{y}_n - Z_n\beta)/\sigma \sim \mathcal{N}(0, h_{n,n})$ using theorem [1.6]. Applying [2.75] and the definition of Student's law, we deduce that:

$$\frac{\widehat{y}_n - Z_n\beta}{\widehat{\sigma}\sqrt{h_{n,n}}} \sim T_{N-r}$$

where T_{N-r} denotes a Student r.v. with $(N-r)$ degrees of freedom.

To provide a confidence interval for $Z_n\beta$, this expression can be rewritten as:

$$\widehat{y}_n - \widehat{\sigma}\delta(\alpha) \leq Z_n\beta \leq \widehat{y}_n + \widehat{\sigma}\delta(\alpha) \qquad [2.77]$$

with $\delta(\alpha) = T_{N-r}^{[-1]}(1-\alpha/2)\sqrt{h_{n,n}}$

where $\widehat{\sigma}$ is derived from [2.76].

If $h_{n,n}$ is small, i.e. close to $1/N$, see [2.56], then $\widehat{y}_n$ is weakly dispersed around the true value $Z_n\beta$. On the other hand, if $h_{n,n} \approx 1$, then $\widehat{y}_n$ is strongly dispersed around the true value $Z_n\beta$. Hence, the term leverages for $h_{n,n}$.

2.6.3.8. *Calculating a confidence interval for σ^2*

The distribution of $\widehat{\sigma}^2/\sigma^2$ is the sum of the squares of $(N-r)$ independent, Gaussian r.v.s of variance 1. $\widehat{\sigma}^2/\sigma^2$ is thus distributed as a χ^2 distribution with $(N-r)$ degrees of freedom, which can be written as:

$$\frac{\widehat{\sigma}^2}{\sigma^2} \sim \chi^2_{N-r} \qquad [2.78]$$

From this, we may deduce a confidence interval of level $100(1-\alpha)\%$ of σ, which can be written as:

$$\frac{\widehat{\sigma}}{\sqrt{\chi^{2[-1]}_{N-r}(1-\alpha/2)}} \leq \sigma \leq \frac{\widehat{\sigma}}{\sqrt{\chi^{2[-1]}_{N-r}(\alpha/2)}} \qquad [2.79]$$

2.6.3.9. *Calculating a confidence interval on "fresh" observations*

Using the estimated value of β for N learning data points $(y_{n=0...N-1}, Z_{n=0...N-1})$, we can deduce the prediction of a "fresh" (new) observation z_o, which is a vector of length $(p+1)$ beginning by 1. This can be written as:

$$\widehat{y}_o = z_o^T \widehat{\beta} \qquad [2.80]$$

The confidence interval at $100(1-\alpha)\%$ for y_o can be written as:

$$\widehat{y}_o - \widehat{\sigma}\delta(\alpha) \leq y_o \leq \widehat{y}_o + \widehat{\sigma}\delta(\alpha) \qquad [2.81]$$

with $\delta(\alpha) = T_{N-r}^{[-1]}(1-\alpha/2)\sqrt{1 + z_o^T(Z^TZ)^{-1}z_o}$

where $\widehat{\sigma}$ is derived from [2.76].

PROOF.— From [2.80], we get $y_o - \widehat{y}_o = z_o^T(\beta - \widehat{\beta}) + \sigma\epsilon_o$. Assuming that the design matrix Z is full column rank, we derive that $\widehat{y}_o$ is a centered Gaussian r.v. with variance:

$$\text{var}(\widehat{y}_o) = \sigma^2(z_o^t(Z^tZ)^{-1}z_o + 1) \qquad [2.82]$$

Using [2.60] we can write $\frac{y_o - \widehat{y}_o}{\sigma} = -z_o^T(Z^TZ)^{-1}Z^T\epsilon + \epsilon_o$. Following [2.75], the r.v. $(y_o - \widehat{y}_o)/\sigma$ is independent from $\widehat{e}/\sigma$. Hence taking the ratio we have:

$$\frac{y_o - \widehat{y}_o}{\widehat{\sigma}\sqrt{1 + z_o^T(Z^TZ)^{-1}z_o}} \sim T_{N-r}$$

■

Note that the confidence interval of $Z_n\beta$ given by [2.77] is narrower than the confidence interval of $\widehat{y}_o$ given by [2.81]. This is legitimate, given that it is harder to make a prediction using new data points than using data taken from the learning set (see exercise 2.13).

2.6.3.10. *Testing the hypothesis $H_0 = \{\beta_k = 0\}$*

We will begin by showing that:

$$\frac{\widehat{\beta}_k - \beta_k}{\widehat{\sigma}\sqrt{[(Z^TZ)^{-1}]_{k,k}}} \sim T_{N-p-1} \qquad [2.83]$$

where T_{N-p-1} denotes a Student r.v. with $(N-p-1)$ degrees of freedom.

PROOF.— Following [2.62], in the case where Z is of full column rank:

$$\frac{\widehat{\beta} - \beta}{\sigma} \sim \mathcal{N}(0; (Z^TZ)^{-1})$$

Multiplying the left-hand side by the vector, all components of which are null except the kth component, which is 1, and then dividing by $\sqrt{[(Z^T Z)^{-1}]_{k,k}}$, we obtain:

$$\frac{\widehat{\beta}_k - \beta_k}{\sigma \sqrt{[(Z^T Z)^{-1}]_{k,k}}} \sim \mathcal{N}(0;1) \qquad [2.84]$$

Dividing the first member of [2.84] by the square root of the first member of [2.78], we obtain [2.83].

It is worth noting that expression [2.83] is not a statistic, as it depends on the unknown parameter β. Such a function is said to be *pivotal*. ∎

Expression [2.83] can be used to derive a critical function for testing the hypothesis $H_0 = \{\beta_k = 0\}$. Indeed, under the hypothesis H_0, we have:

$$T_k(y) = \frac{\widehat{\beta}_k}{\widehat{\sigma} \sqrt{[(Z^T Z)^{-1}]_{k,k}}} \sim T_{N-p-1} \qquad [2.85]$$

$T_k(y)$ is also referred to in the literature as the *Z-score*. Roughly speaking, a Z-score larger than two in absolute value is significantly non-zero (at the confidence level of 5%). This provides a method for model selection. By model selection, we mean the selection of significant explanatory variables.

The bilateral test of H_0 with critical function $\mathbb{1}(T_k(y) \notin (-\eta, \eta))$ can therefore be used, where η is determined for a given significance level. Based on expression [2.35], this bilateral test has a p-value of:

$$p\text{-value} = 2 \int_{|T_k|}^{+\infty} p_{T_{N-p-1}}(t) dt \qquad [2.86]$$

where $p_{T_{N-p-1}}(t)$ is the probability density of Student's distribution with $(N-p-1)$ degrees of freedom. As stated in section 2.7, the Student distribution is a specific instance of the Fisher distribution.

2.6.3.11. *Testing the hypothesis $H_0 = \{\beta_1 = \cdots = \beta_p = 0\}$*

We will test the hypothesis $H_0 = \{\beta_1 = \cdots = \beta_p = 0\}$. To do this, we take:

– the mean $y_c = \frac{1}{N} \sum_{n=0}^{N-1} y_n$;

– the explained sum of squares:

$$\text{ESS} = \|\widehat{y} - y_c\|^2 = \sum_{n=0}^{N-1} (\widehat{y}_n - y_c)^2 \qquad [2.87]$$

Following [2.74], the ratio ESS/σ^2 follows, under H_0, a χ^2 distribution with p degrees of freedom;

– the residual sum of squares:

$$\text{RSS} = \widehat{e}^T e = \sum_{n=0}^{N-1} (y_n - \widehat{y}_n)^2 \qquad [2.88]$$

Following [2.75], the ratio RSS/σ^2 follows, under H_0, a χ^2 distribution with $N - (p+1)$ degrees of freedom;

– the total sum of squares:

$$\text{TSS} = \|y - y_c\|^2 = \sum_{n=0}^{N-1} (y_n - y_c)^2 \qquad [2.89]$$

Based on Pythagoras' theorem (see also Figure 2.1), we have $\text{TSS} = \text{RSS} + \text{ESS}$;

– coefficient of determination R^2, such that:

$$R^2 = \frac{\text{ESS}}{\text{TSS}} = 1 - \frac{\text{RSS}}{\text{TSS}} \in (0,1) R^2 \qquad [2.90]$$

The closer the R^2 is to 1, the smaller the residuals, and the better the model will explain observations. Inversely, if $R^2 \approx 0$, there will be little connection between the set of explanatory variables and the response y. This may be used to deduce a test for the hypothesis H_0. To do this, we denote:

$$F = \frac{(N-p-1)R^2}{p(1-R^2)} = \frac{p^{-1}\text{ESS}}{(N-p-1)^{-1}\text{RSS}} \qquad [2.91]$$

In accordance with section 2.7, the statistic F follows, under H_0, a Fisher distribution with $(p, N-p-1)$ degrees of freedom, which can be written as:

$$F \sim F_{p,N-p-1} \qquad [2.92]$$

We can therefore construct for the hypothesis H_0 the critical test function $T = \mathbb{1}(F \notin (-\eta, \eta))$ and calculate the threshold η for a given significance level, along with a p-value.

We also consider the *adjusted* R^2:

$$\text{adjusted } R^2 = 1 - \frac{\text{RSS}/(N-p-1)}{\text{TSS}/(N-1)} \qquad [2.93]$$

Maximizing *adjusted* R^2 is equivalent to minimizing $\text{RSS}/(N-p-1)$, but while RSS always decreases as the number of variables in the model increases, *adjusted* R^2

may increase or decrease, due to the presence of p in the denominator. Therefore, $adjusted\ R^2$ can be used for model selection, as briefly presented in section 2.6.3.13.

In example 2.9, we see that the statistics TSS, ESS, R^2, $adjusted\ R^2$ and F are invariant by standardization, whereas $\widehat{\beta}$ is not.

2.6.3.12. AIC and BIC

As mentioned previously, RSS cannot be used for model selection; indeed, it always decreases as the number of variables in the model increases. A simple idea consists of penalizing the high order as with $adjusted\ R^2$. This leads to two commonly used indices of model quality in statistics: the *Akaike information criterion* (AIC) and the *Bayesian information criterion* (BIC). The expressions for the linear model are given by:

$$\text{AIC} = N \log(\text{RSS}/N) + 2(p+1) \qquad [2.94]$$

$$\text{BIC} = N \log(\text{RSS}/N) + (p+1)\log(N) \qquad [2.95]$$

The preferred model is the one with either the minimum AIC value or the minimum BIC value. For AIC derivation, see exercise 2.19. For a detailed comparison of AIC and BIC, see [BUR 02].

The linear model analysis is summarized as follows:

Data: $y_{n=0:N-1}$, $X_{n=0:N-1} \in \mathbb{R}^p$
begin
    Form the design matrix $Z = \begin{bmatrix} \mathbf{1}_N & X \end{bmatrix}$;
    perform the mean y_c;
    perform $A = (Z^T Z)^{-1}$;
    perform $\Pi_Z = ZAZ^T$, leverage $h_{n,n} = \Pi_{Z,n,n}$;
    perform $\widehat{\beta} = AZ^T y$, $\widehat{y} = Z\widehat{\beta}$ and $\widehat{e} = y - \widehat{y}$;
    perform $\widehat{\sigma}^2 = \frac{1}{N-p-1}\widehat{e}^T\widehat{e} \sim \sigma^2 \times \xi^2_{N-p-1}$;
    perform Z_k–score $= \dfrac{\widehat{\beta}_k}{\widehat{\sigma}\sqrt{A_{k,k}}} \sim T_{N-p-1}$ for $k = 0:p$;
    perform the confidence interval on $\widehat{\beta}$ as [2.83];
    perform the confidence interval on "fresh" data as [2.80];
    perform ESS $= \|\widehat{y} - y_c\|^2$, RSS $= \widehat{e}^T e$, TSS $= \|y - y_c\|^2$;
    perform $F = \dfrac{p^{-1}\text{ESS}}{(N-p-1)^{-1}\text{RSS}} \sim F_{p,N-p-1}$;
    perform $R^2 = \dfrac{\text{ESS}}{\text{TSS}}$;
    perform $adjusted\ R^2 = 1 - \dfrac{\text{RSS}/(N-p-1)}{\text{TSS}/(N-1)}$;
    perform AIC $= N \log(\text{RSS}/N) + 2(p+1)$;
    perform BIC $= N \log(\text{RSS}/N) + (p+1)\log(N)$;
end
/* Z is assumed to be full rank column $p+1$ */

Algorithm 3: Linear regression calculation

EXAMPLE 2.8 (Mean estimation).– Mean estimation may be seen via linear model writing, with obvious notations, $y = m\mathbf{1} + \sigma\epsilon$. Determine an estimator of $\hat{m}$ and an estimator of σ. Determine the distribution of these estimators.

HINTS: *Therefore, $Z = \mathbf{1}$, $Z^T Z = N$ and $\hat{m} = N^{-1}\sum_{n=0}^{N-1} X_n$. Following [2.62], $\hat{m}$ is unbiased. Following [2.76], $\hat{\sigma}^2 = (N-1)^{-1}\sum_{n=0}^{N-1}(X_n - \hat{m})^2$, which is unbiased. Also, $\hat{m}$ and $\hat{\sigma}^2$ are independent.*

Following [2.83], $(m - \hat{m})\sqrt{N}/\hat{\sigma} \sim T_{N-1}$. A confidence interval of m at 95% is written as:

$$\hat{m} - \frac{\hat{\sigma}}{\sqrt{N}}T^{[-1]}_{N-1}(1-\alpha/2) \leq m \leq \hat{m} + \frac{\hat{\sigma}}{\sqrt{N}}T^{[-1]}_{N-1}(1-\alpha/2)$$

with $T^{[-1]}_{N-1}(1-\alpha/2) = 2.01$ for $N = 50$. This result can be compared with the confidence interval, when σ is known, which is a little smaller (see example 2.1).

Because $\hat{\sigma}$ tends to σ, the asymptotic variance of $\hat{m}$ is σ^2/N, meaning that $\hat{m}$ is asymptotically efficient (see example 2.6).

■

EXAMPLE 2.9 (Water fluoridation).– The values reported in Table 2.2 give the occurrence of dental cavities and the level of fluoride in drinking water observed in 21 American cities in a total of 7257 children [MC 43]. These values are illustrated on the left-hand side of Figure 2.9. The linear model does not appear to be suitable. A model of the form $y \approx \gamma(x+1)^\alpha$ with $\alpha < 0$ seems to give a more satisfactory result. Taking the logarithm of the two members, we obtain $\log y \approx \beta_0 + \beta_1 \log(x+1)$, which is linear with respect to $\log(x+1)$. This corresponds well with the illustration on the right-hand side of Figure 2.9.

fluoride	1.90	2.60	1.80	1.20	1.20	1.20	0
cavities	17.13	17.85	18.29	18.72	20.39	21.99	23.44
fluoride	1.30	0.90	0.60	0.50	0.40	0.30	0
cavities	24.89	29.90	32.22	40.35	47.32	51.02	51.23
fluoride	0.20	0.10	0.20	0.10	0.10	0.10	0
cavities	53.19	56.02	59.73	75.26	58.78	48.84	52.40

Table 2.2. *Cavity values given in thousands. The fluoride level is given in ppm (parts per million)*

Take $y = \beta_0 + \beta_1 x$, where x denotes the log-percent of the number of cavities and y the log-concentration of fluoride. Write a program to perform the statistics RSS, TSS, R^2, adjusted R^2 and F, and test the hypothesis $\beta_1 = 0$.

78 Digital Signal Processing with Python Programming

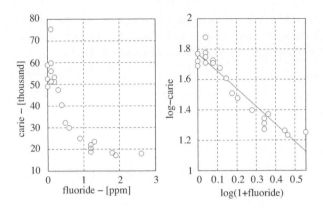

Figure 2.9. *Percentage of dental cavities observed as a function of fluoride levels in drinking water in ppm (parts per million)*

HINTS: *Run the following program:*

```
# -*- coding: utf-8 -*-
"""
Created on Wed Jun  8 08:57:15 2016
****** testfluorcaries
@author: maurice
"""
from numpy import array, log10, ones, sum, mean, dot, zeros, std
from numpy.linalg import pinv
from scipy.stats import f
import scipy.stats as ss
caries = array([\
    1.90,2.60,1.80,1.20,1.20,1.20,\
    1.30,0.90,0.60,0.50,0.40,0.30,\
    0.20,0.10,0.20,0.10,0.10,0.10,0,0,0]);
fluor = array([17.13,17.85,18.29,18.72,20.39,21.99,23.44,\
    24.89,29.90,32.22,40.35,47.32,51.02,51.23,\
    53.19,56.02,59.73,75.26,58.78,48.84,52.40]);
xns = log10(caries+1);
# to see effect of the standardization, test 0 and 1
standardizatioflag = True
if standardizatioflag:
    xc = xns-mean(xns)
    x = xc / std(xns)
else:
```

```
x = xns
y = log10(fluor);
N = len(x); H =zeros([N,2])
H[:,0] = ones(N); H[:,1] = x; invH = pinv(H)
hatbeta = dot(invH,y);
haty = dot(H,hatbeta);
ESS = sum((haty-mean(y)) **2); TSS = sum((y-mean(y)) **2);
RSS = TSS-ESS; R2 = ESS/TSS;
adjR2 = 1.0-(RSS/(N-2.0)/(TSS/(N-1.0)))
F = (N-2)*R2/(1.0-R2);
pvalue = 1-f.cdf(F,1,N-2);
print('\t******* p-value = %3.1e\n'%pvalue)
print(hatbeta, ESS, TSS, R2, adjR2, F, pvalue)
```

We obtain $R^2 = 0.9092$ and $F = \frac{(N-2)R^2}{(1-R^2)} = 190.20$, and the p-value $\approx 2.4 \times 10^{-11}$. This leads us to reject the hypothesis $\beta_1 = 0$.

Assigning the value True to the variable standardiseflag, we note that $SST, R2, S, pvalue$ are invariant by standardization, whereas beta is not. ∎

EXERCISE 2.13 (Confidence intervals on linear model features).– (see p. 176) The file statsmodels.api.datasets.diabetes.load() consists of 10 explanatory variables (age, sex, body mass index, blood pressure and 6 blood serum measurements) measured on 442 patients, and an indication of disease progression after one year. Let us verify that all explanatory variable values have zero mean and norm 1.

We denote y_n as the indication of disease progression and x_n as the age, and write $y_n = \beta_0 + \beta_1 x_n + \sigma \epsilon_n$.

Write a program to perform the regression coefficients β, the predictions $\widehat{y}_n$, the confidence intervals of $Z_n \beta$, expression [2.77] and, for an arbitrary fresh values, the confidence interval of $\widehat{y}_o$, expression by [2.81].

EXERCISE 2.14 (Hypothesis test on $H_0 = \{\beta_{1:p} = 0\}$).– (see p. 177) The file sklearn.datasets.load_boston() consists of the Boston housing price (median value of owner-occupied homes in $1000's) considered as the target data and 13 features considered as the explanatory variables. We consider that the target can be modeled by $y_n = \beta_0 + \sum_{k=1}^{p} \beta_k x_{n,k} + \sigma \epsilon_n$, where $x_{n,k}$ are the following selected explanatory variables: RM, LSTAT, CRIM, ZN, CHAS, DIS (6 among 13). Determine the value of the statistic F, equation [2.91], and the p-value associated with the test of the hypothesis $H_0 = \{\beta_1 = \ldots = \beta_p = 0\}$.

EXERCISE 2.15 (Model selection based on Z-score).– (see p. 178) Let us consider the file statsmodels.api.datasets.diabetes.load() of exercise 2.13. We let

$y_n = \beta_0 + \sum_{k=1}^{10} \beta_k x_{n,k} + \sigma \epsilon_n$, where $x_{n,k}$ denotes the kth features associated with the patient n. Write a program to perform the Z-score [2.85] for the intercept and for the 10 explanatory variables. Conclude.

2.6.3.13. *Feature selection in the linear model*

An important concern is to select only the relevant explanatory variables in the linear model. *Brute force* consists of computing a significative value as, e.g. the *adjusted* R^2, for all the possible subsets of explanatory variables, i.e. for the 2^p combinations, but for large p, this computation could be cumbersome.

Here we only present the approach called *backward stepwise selection*. This consists of two steps: in the first step, we start with the full linear model consisting of all p predictors, and then iteratively remove the least useful predictor, one at a time. This leads to $p+1$ models. Let us denote $\mathcal{M}_0, \ldots, \mathcal{M}_p$ as these models, where $\mathcal{M}_k$ is specified by k explanatory variables and the response y. In second step, we select the best among them and use a criterion, e.g. AIC, BIC or *adjusted* R^2, etc. or also a cross-validation-type approach, as presented in section 2.6.8. For more details, see [HAS 09].

This procedure is summarized in the following algorithm 4:

Data: $X \in \mathbb{R}^p \times \mathbb{R}^N, y \in \mathbb{R}^N$, with $N > p$
Result: best model
begin
    for $k = p$ *to* 1, *step* -1 **do**
        for $j = 1$ *to* k **do**
            compute T_j of the model $\mathcal{M}_{k,j}$;
        end
        $T^* = \arg\min_j T_j$;
        $\mathcal{M}_k = \mathcal{M}_{k,T^*}$;
    end
    Select the best model from among $\mathcal{M}_0, \ldots, \mathcal{M}_p$ using selection criterium;
end
/* T_j denotes a scalar statistic, as e.g. the norm of the prediction residue */
/* A model $\mathcal{M}_k$ is specified by a list of explanatory variables and the target y */
/* The model selection criterium could be AIC, BIC, etc, or cross-validation */

Algorithm 4: Backward stepwise selection

EXERCISE 2.16 (Model selection based on *adjusted* R^2, AIC and BIC).– (see p. 178) Let us consider the file `statsmodels.api.datasets.diabetes.load()` of exercise 2.13. We let $y_n = \beta_0 + \sum_{k=1}^{10} \beta_k x_{n,k} + \sigma \epsilon_n$, where $x_{n,k}$ denotes the kth features associated with the patient n.

Write a program which implements the backward stepwise selection algorithm 4, successively using the *adjusted* R^2, the AIC and the BIC as the model selection criterium. Compare the three selected models.

2.6.3.14. *Weighted least squares (WLS)*

In some cases, for the model $y = Z\beta + \sigma\epsilon$, the covariance of ϵ is a known matrix $\Gamma \neq I$. The best estimator in terms of least squares can then be expressed as:

$$\widehat{\beta} = (Z^T \Gamma^{-1} Z)^{-1} Z^T \Gamma^{-1} Y, \qquad [2.96]$$

[2.96] is known as the weighted least squares (WLS) estimator.

PROOF.– Let P be the square matrix of dimension N, such that $PP^T = \Gamma^{-1}$. We have $Py = (PZ)\beta + \sigma\xi$, with $\xi = P\epsilon$. We can verify that $\text{cov}(\xi) = I_N$, as in the case of ordinary least squares (OLS). Therefore, we can use the expression [2.60], replacing y by Py and Z by PZ. From this, we deduce that the estimator is given by [2.96]. Moreover, it is unbiased, and its covariance is expressed as:

$$\text{cov}\left(\widehat{\beta}\right) = \sigma^2 (Z^T \Gamma^{-1} Z)^{-1} \qquad [2.97]$$

It is easy to show that the weighted least squares estimator minimizes the following function:

$$J(\beta) = (Y - Z\beta)^T \Gamma^{-1} (Y - Z\beta) = \|y - \beta Z\|_{\Gamma^{-1}}^2 \qquad [2.98]$$

Hence, given that $\Gamma \geq 0$, we can derive [2.96] directly from the projection theorem. ∎

2.6.4. *Method of moments*

Let us consider N observations $X_0, \ldots, X_{N-1}$ i.i.d. from a Gamma distribution of parameter $\theta = (k, \lambda) \in \mathbb{R}^+ \times \mathbb{R}^+$, with the probability density:

$$p_X(x; \theta) = \frac{1}{\lambda^k \Gamma(k)} x^{k-1} e^{-x/\lambda} \mathbb{1}(x \geq 0) \qquad [2.99]$$

where $\Gamma(k) = \int_0^{+\infty} t^{k-1} e^{-t} dt$. We can show that $\mathbb{E}\{X_n\} = k\lambda$ and $\mathbb{E}\{X_n^2\} = k(1+k)\lambda^2$.

The basic idea behind the method of moments is to relate statistical and empirical moments, which can be written as:

$$\{\{2 \quad \begin{array}{l} k\lambda \approx \frac{1}{N} \sum_{n=0}^{N-1} X_n \\ k(1+k)/\lambda^2 \approx \frac{1}{N} \sum_{n=0}^{N-1} X_n^2 \end{array}$$

To obtain an estimator of θ, we must simply solve the above two equations with respect to the two unknown parameters (k, λ). Hence:

$$\begin{cases} \widehat{k} = \dfrac{\widehat{m}^2}{\widehat{\sigma}^2} \\ \widehat{\lambda} = \dfrac{\widehat{\sigma}^2}{\widehat{m}} \end{cases} \qquad [2.100]$$

where $\widehat{m} = N^{-1} \sum_{n=0}^{N-1} X_n$ and $\widehat{\sigma}^2 = N^{-1} \sum_{n=0}^{N-1} (X_n - \widehat{m})^2$.

EXERCISE 2.17 (Moment estimator: central limit theorem).– (see p. 180) For the moment estimator given by [2.100], determine the asymptotic behavior deduced from the central limit theorem 2.9 and the continuity theorem 2.10.

2.6.4.1. *Generalized method of moments*

Clearly, multiple estimators may be obtained by choosing moments of the form $\mathbb{E}_\theta \{g(X_n)\}$. It is even possible to use more moments then parameters to estimate, leading to more equations than unknowns. Starting with M moments of the form $\mathbb{E}\{g_m(X_0)\} = S_m(\theta)$ (X_0 may be used with no loss of generality, as the r.v.s are considered to be i.i.d.), the generalized method of moments consists of taking the following estimator:

$$\widehat{\theta} = \arg\min_\theta \sum_{m=1}^{M} \left(\frac{1}{N} \sum_{n=0}^{N-1} g_m(X_n) - S_m(\theta) \right)^2$$

$$= \arg\min_\theta \|\widehat{S} - S(\theta)\|^2 \qquad [2.101]$$

where $\widehat{S}$ is a vector of length M, written as:

$$\widehat{S} = \begin{bmatrix} \frac{1}{N} \sum_{n=0}^{N-1} g_0(X_n) \\ \vdots \\ \frac{1}{N} \sum_{n=0}^{N-1} g_{M-1}(X_n) \end{bmatrix} \quad \text{and} \quad S(\theta) = \begin{bmatrix} \mathbb{E}_\theta \{g_0(X_0)\} \\ \vdots \\ \mathbb{E}_\theta \{g_{M-1}(X_0)\} \end{bmatrix} \qquad [2.102]$$

The estimator given by expression [2.101] may be improved by using the covariance matrix of $\widehat{S}$ and by using the weighted least squares approach given by expression [2.98]. This yields the following estimator:

$$\hat{\theta} = \arg\min_{\theta}(\widehat{S} - S(\theta))^T C^{-1}(\theta)(\widehat{S} - S(\theta)) \qquad [2.103]$$

where the covariance matrix is given by:

$$C(\theta) = \mathbb{E}_\theta\left\{(\widehat{S} - \mu(\theta))(\widehat{S} - \mu(\theta))^T\right\}, \quad \text{where} \quad \mu(\theta) = \mathbb{E}_\theta\left\{\widehat{S}\right\} \qquad [2.104]$$

Using the fact that the r.v.s X_n are i.i.d., the (m, m')-entry of $C(\theta)$ is written as:

$$C_{m,m'}(\theta) = \mathbb{E}_\theta\left\{(g_m(X_0) - \mu_m(\theta))(g_{m'}(X_0) - \mu_{m'}(\theta))\right\}$$
$$= \mathbb{E}_\theta\left\{g_m(X_0) g_{m'}(X_0)\right\} - \mu_m(\theta)\mu_{m'}(\theta)$$

[2.103] is generally minimized using a computational approach via a minimization program such as `scipy.optimize.fmin`.

When $C(\theta)$ has no closed form expression, another possibility is to replace the matrix $C(\theta)$ in expression [2.103] with a consistent estimation $\widehat{C}$, giving:

$$\hat{\theta} = \arg\min_{\theta}(\widehat{S} - S(\theta))^T \widehat{C}^{-1}(\widehat{S} - S(\theta)) \qquad [2.105]$$

then to apply a general minimization program.

A major problem with the generalized method of moments is that there is nothing to indicate, *a priori*, how many and which moments we have to choose, except in cases where a "sufficient" statistic exists (see the Fisher factorization criterion in [NEY 33, STI 73]). However, in these cases, the maximum likelihood approach is preferable.

In conclusion, the method of moments is easy to apply, but a case-by-case study is required in order to obtain satisfactory performances. However, in the case of large samples, the law of large numbers 2.8 and the central limit theorem 2.9 can be used for performance evaluation.

EXERCISE 2.18 (Moment estimators of mixture proportion).– (see p. 181)
Consider a sequence of N random variables X_n i.i.d., where the common distribution has the following probability density:

$$p_X(x) = \frac{\alpha}{\sigma_0\sqrt{2\pi}} e^{-(x-m_0)^2/2\sigma_0^2} + \frac{1-\alpha}{\sigma_1\sqrt{2\pi}} e^{-(x-m_1)^2/2\sigma_1^2}$$

where $\alpha \in (0, 1)$. We suppose that α is unknown, but m_1, m_2, σ_1 and σ_2 are known:

1) Give the expression of the statistical model. Use this result to deduce the expressions of $\mathbb{E}\{X_n\}$. Derive an estimator based on $\mathbb{E}\{X_n\}$ and $\mathbb{E}\{X_n^2\}$.

2) Determine a moment estimator for the parameter α, which is based on the statistic $\widehat{S}(X) = N^{-1} \sum_{n=0}^{N-1} X_n$.

3) Determine a moment estimator for α based on two statistics, $\widehat{S}_0(X) = N^{-1} \sum_{n=0}^{N-1} X_n$ and $\widehat{S}_1(X) = N^{-1} \sum_{n=0}^{N-1} X_n^2$, using the expression of $S(\alpha)$ given by [2.102].

4) Determine the expression of the covariance $C(\alpha)$, defined by [2.104]. Use expression [2.103] to find a moment estimator for α.

5) Write a program to simulate and compare the behavior of the three estimators.

2.6.5. *Maximum likelihood approach*

The maximum likelihood estimation method is based on the fact that a good estimator of θ may be reasonably thought to maximize the probability of what has been observed. In the case of "continuous" random variables, the probability is replaced by the probability density.

DEFINITION 2.10.– Let us consider a parametric model with a likelihood $\theta \mapsto p_X(x; \theta)$. A maximum likelihood estimator (MLE) is an estimator defined by:

$$\widehat{\theta}_{\mathrm{MLE}} = \arg\max_{\theta \in \Theta} p_X(X; \theta) \qquad [2.106]$$

As the logarithm is an increasing function, the likelihood may be replaced by the log-likelihood in expression [2.106]. Note that:

– one sufficient condition for the existence of a maximum is that the set Θ is compact and the function defined by [2.13] is continuous over Θ;

– the MLE is *not* generally unique;

– the MLE is invariable by re-parameterization. This means that if $\theta \mapsto g(\theta)$ is a function defined over Θ and if $\widehat{\theta}$ is an MLE of θ, then $g(\widehat{\theta})$ is an MLE of $g(\theta)$.

In most practical situations, the maximum likelihood estimator converges toward the true value of the parameter as the number of samples tends toward infinity. More precisely, under very general conditions, it can be shown that:

$$\sqrt{N}\left(\widehat{\theta}_{\mathrm{MLE}} - \theta\right) \xrightarrow{d} \mathcal{N}(0, C) \qquad [2.107]$$

where

$$C = \lim_{N \to +\infty} N F^{-1} \quad \text{with} \quad F = -\mathbb{E}_\theta \left\{ \partial_\theta^2 \log p(X_{0:N-1}; \theta) \right\} \quad [2.108]$$

F is the Fisher information matrix, as given by expression [2.41].

This result is fundamental, showing that maximum likelihood estimators are asymptotically unbiased and asymptotically efficient, in that their asymptotic covariance matrix is the limit of the Cramer-Rao bound, as given by expression [2.43].

This is the reason why this estimator is generally preferred, even when problem [2.106] can only be solved using a computational approach. Note, however, that the maximum likelihood approach can fail: an example is shown in exercise 2.25.

When $X_0, \ldots, X_{N-1}$ is a sequence of N i.i.d. r.v.:

$$C = F^{-1} \quad \text{with} \quad F = -\mathbb{E}_\theta \left\{ \partial_\theta^2 \log p(X_0; \theta) \right\} \quad [2.109]$$

EXAMPLE 2.10 (Poisson distribution).– For discrete observations, the MLE can be extended using the probability instead of the probability density. Let us consider N observations i.i.d., with values in $\mathbb{N}$ and a Poisson distribution with parameter λ, which can be written as:

$$\mathbb{P}\{X_n = x\} = \frac{\lambda^x}{x!} e^{-\lambda}$$

where $\lambda \in \mathbb{R}^+$. Determine an MLE for λ.

HINTS: *The log-likelihood is written as:*

$$\ell(\lambda) = -N\lambda + \log \lambda \sum_{n=0}^{N-1} X_n - \sum_{n=0}^{N-1} X_n!$$

Canceling the derivative with respect to λ, we obtain $\widehat{\lambda} = N^{-1} \sum_{n=0}^{N-1} X_n$. It is worth noting that the r.v. $\sum_{n=0}^{N-1} X_n$ is Poisson distributed with parameter $N\lambda$, but the distribution of $\widehat{\lambda}$ is very complex. Using the central limit theorem 1.9, it can be approximated by a Gaussian distribution with mean λ and variance λ/N. From

$$\partial_\lambda^2 \ell(X) = -\frac{\sum_{n=0}^{N-1} X_{kn}}{\lambda^2} \Rightarrow \mathbb{E}_\lambda \left\{ \partial_\lambda^2 \ell(X) \right\} = -\frac{N}{\lambda}$$

we derive the Cramer-Rao bound λ/N, which means that the estimator is efficient. ∎

2.6.5.1. *MLE of the i.i.d. Gaussian model*

Consider a sequence of N Gaussian observations { i.i.d. $\mathcal{N}(N; m, C)$}, where m is a vector of dimension d and C is a square matrix of dimension d, assumed to be strictly positive. C is therefore invertible with $\det\{C\} \neq 0$. The log-likelihood is written as:

$$\ell(\theta) = -\frac{Nd}{2}\log(2\pi) - \frac{N}{2}\log\det\{C\}$$
$$-\frac{1}{2}\sum_{k=0}^{N-1}(X_k - m)^T C^{-1}(X_k - m) \quad [2.110]$$

where $\theta = (m, C) \in \mathbb{R}^d \times \mathcal{M}_d^+$. The maximum likelihood estimators of the i.i.d. Gaussian model are:

$$\widehat{m}_{\text{MLE}} = \frac{1}{N}\sum_{k=0}^{N-1} X_k \quad [2.111]$$

$$\widehat{C}_{\text{MLE}} = \frac{1}{N}\sum_{k=0}^{N-1}(X_k - \widehat{m}_{\text{MLE}})(X_k - \widehat{m}_{\text{MLE}})^T \quad [2.112]$$

and the likelihood maximum is expressed as:

$$\ell_{\max} = -\frac{Nd}{2}\log(2\pi) - \frac{N}{2}\log\det\{\widehat{C}_{\text{MLE}}\} - \frac{Nd}{2} \quad [2.113]$$

PROOF.– Maximization of ℓ with respect to m is carried out by canceling the gradient of ℓ:

$$\partial_m \ell(\theta) = \sum_{k=0}^{N-1} C^{-1}(X_k - m) = 0$$

Hence, assuming that C is of full rank, we have:

$$\widehat{m}_{\text{MLE}} = \frac{1}{N}\sum_{k=0}^{N-1} X_k$$

Let:

$$\widehat{C} = \frac{1}{N}\sum_{k=0}^{N-1}(X_k - \widehat{m}_{\text{MLE}})(X_k - \widehat{m}_{\text{MLE}})^T$$

Substituting $\widehat{m}_{\mathrm{MLE}}$ and $\widehat{C}$ into ℓ, using the identity $v^T A v = \operatorname{trace}\{A v v^T\}$ and the linearity of the trace, we obtain the log-likelihood as a function of C:

$$\widetilde{\ell}(C) = -\frac{Nd}{2}\log(2\pi) - \frac{N}{2}\log\det\{C\} - \frac{N}{2}\operatorname{trace}\{C^{-1}\widehat{C}\} \qquad [2.114]$$

We now need to maximize $\widetilde{\ell}(C)$ with respect to C. To do this, the trace and the determinant (for matrices of ad hoc dimensions) verify:

$$\operatorname{trace}\{AB\} = \operatorname{trace}\{BA\} \text{ and } \det\{AB\} = \det\{BA\} = \det\{A\}\det\{B\}$$

Let:

$$S = \widehat{C}^{-1/2} C \widehat{C}^{-1/2} \qquad [2.115]$$

S is positive and can therefore be diagonalized. Substituting [2.115] into [2.114], we then obtain:

$$\begin{aligned}\widetilde{\ell}(C) &= -\frac{Nd}{2}\log(2\pi) - \frac{N}{2}\log\det\{\widehat{C}^{1/2} S \widehat{C}^{1/2}\} - \frac{N}{2}\operatorname{trace}\{S^{-1}\} \\ &= -\frac{Nd}{2}\log(2\pi) - \frac{N}{2}\log\det\{\widehat{C}\} - \frac{N}{2}\log\det\{S\} - \frac{N}{2}\operatorname{trace}\{S^{-1}\}\end{aligned}$$
$$[2.116]$$

Let λ_s be the eigenvalues of S. Expression [2.116] can be rewritten as:

$$\widetilde{\ell}(C) = -\frac{Nd}{2}\log(2\pi) - \frac{N}{2}\log\det\{\widehat{C}\} - \frac{N}{2}\sum_{s=0}^{d-1}\left(\log\lambda_s + \lambda_s^{-1}\right) \qquad [2.117]$$

Canceling the derivative with respect to λ_s, we obtain $\lambda_s = 1$, and thus $S = I_d$ which is clearly a positive matrix. Using $S = I_d$ in expression [2.115], we deduce that the matrix C that maximizes the likelihood is, in fact, the matrix $\widehat{C}$. ■

In the specific case where $d = 1$, we obtain:

$$\widehat{m}_{\mathrm{MLE}} = \frac{1}{N}\sum_{k=0}^{N-1} X_k$$

$$\widetilde{\sigma}_{\mathrm{MLE}}^2 = \frac{1}{N}\sum_{k=0}^{N-1} (X_k - \widehat{m}_{\mathrm{MLE}})^2$$

$$\ell_{\max} = -\frac{N}{2}\log(2\pi) - \frac{N}{2}\log\det\{\widehat{\sigma}_{\mathrm{MLE}}^2\} - \frac{N}{2}$$

EXERCISE 2.19 (MLE for the linear model).– (see p. 182) The Gaussian linear model is the statistical model defined by N independent, Gaussian random variables with respective mean $Z_n \theta$ and variance σ^2, where Z_n is given by [2.50]. Let us note that the model is not identically distributed because the mean does depend on n. Determine the MLE of (θ, σ^2).

2.6.5.2. *Voronoi regions*

We consider a likelihood function $p(x; \theta)$, where $x \in \mathbb{R}^p$ and θ belongs to a finite discrete set of values $\Theta = \{\theta_0, \ldots, \theta_{g-1}\}$. The estimation of θ is clearly a classification problem, which consists of determining a function that associates with x, an index value in Θ. This is equivalent to defining a g-partition[3] of $\mathbb{R}^p$ as:

$$\Lambda_i = \{x \in \mathbb{R}^p : p(x; \theta_i) \geq p(x; \theta_j),\ \forall j \neq i\} \quad [2.118]$$

Regions Λ_i are called *Voronoi regions*. They can be used for classification as follows: given a "fresh data" x_o, if x_o belongs to Λ_i, then it can be decided that x_o belongs to class i.

More generally we may replace $p(x; \theta_i)$ by any other metric, as for example, the Mahalanobis distance:

$$d^2(x) = (x - m)^T C^{-1} (x - m) \quad [2.119]$$

where C is a positive square matrix of size p and m is a vector in $\mathbb{R}^p$.

EXERCISE 2.20 (Iris classification).– (see p. 183) Let us consider the iris dataset. For each of the three classes, we do a partition in one training and one testing dataset. We apply to the training dataset the LDA dimensionality reduction determined in exercise 2.1. We assume that the three "reduced" classes may be modeled as Gaussian distributions with respective means m_0, m_1 and m_2 and respective covariances C_0, C_1 and C_2.

Write a program to (i) reduce the feature dimension, (ii) estimate the parameters of the three classes of the model using the training dataset and then after projection in the reduced space, (iii) determine the maximum likelihood class of the testing dataset elements.

EXERCISE 2.21 (Asymptotic distribution of a correlation MLE).– (see p. 184) We consider N i.i.d. bivariate Gaussian observations distributed as $\{$ i.i.d. $\mathcal{N}(N; m, C)\}$. The MLE of the covariance matrix C is given by [2.112]. Let us consider the case $d = 2$ and assume that the correlation is defined by $\rho = C_{0,1}/\sqrt{C_{0,0} C_{1,1}}$.

---

[3] Strictly speaking, [2.118] does not define a partition due to the presence of common borders. But this problem can be avoided by using the strict inequality for $j > i$ and the large inequality for $j < i$.

1) Using [2.112], show that a MLE of ρ is given by:

$$\widehat{\rho} = \frac{\widehat{C}_{0,1}}{\sqrt{\widehat{C}_{0,0}\widehat{C}_{1,1}}} \qquad [2.120]$$

2) Based on the CRB expression [2.47], the asymptotic distributions of $\widehat{\sigma}_1$, $\widehat{\sigma}_2$ and $\widehat{\rho}$ have the standard deviations $\sigma_1/\sqrt{2N}$, $\sigma_2/\sqrt{2N}$ and $(1-\rho^2)/\sqrt{N}$, respectively;

3) Perform a simulation that confirms these results.

EXERCISE 2.22 (MME versus MLE of the correlation).– (see p. 184) We consider N i.i.d. bivariate Gaussian observations distributed as $\{$ i.i.d. $\mathcal{N}(N;0,C)\}$. Let us consider the case $d=2$ and

$$C = \begin{bmatrix} 1 & \rho \\ \rho & 1 \end{bmatrix}$$

with $\rho \in (-1,1)$. We denote $X_{n,0}$ and $X_{n,1}$ as the two components of the bivariate process.

1) Give the expression of the statistical model;

2) Given that the correlation is defined by $\rho = C_{0,1}/\sqrt{C_{0,0}C_{1,1}}$, derive a moment method estimator (MME) of ρ;

3) Derive an MLE of ρ;

4) Perform a simulation program to compare the two estimators.

EXERCISE 2.23 (Correlation GLRT).– (see p. 185) Let us consider two samples $\{X_{0,0}, \ldots, X_{n-1,0}\}$ and $\{X_{0,1}, \ldots, X_{n-1,1}\}$ i.i.d. Gaussian, with respective means m_0 and m_1 and respective variances σ_0^2 and σ_1^2, and with a correlation coefficient $\rho \in (-1,1)$.

1) Based on the results of section 2.6.5.1, particularly on equation [2.113], derive the GLRT for the hypothesis $H_0 = \{|\rho| \leq \rho_0\}$, which tests whether the modulus of the correlation is lower than a given *positive* value ρ_0;

2) Transformation $r \mapsto f = 0.5\log(\frac{1+r}{1-r}) = \arg\tanh(r)$ is known as Fisher's transformation[4]. The Fisher transformations of $\widehat{\rho}$ and ρ_0 are denoted by $\widehat{f}$ and f_0, respectively. Under the hypothesis H_0, the random variable $\widehat{f}$ can be shown to approximately follow a Gaussian distribution of mean f_0 and variance $1/(n-3)$ [SHA 08];

3) Use the previous result to determine the p-value of the test for the hypothesis $H_0 = \{|\rho| =\leq 0.7\}$ using the data provided in Table 2.3.

---

4 arg tanh is the inverse hyperbolic tangent function.

H (cm)	162	167	167	159	172	172	168
W (kg)	48.3	50.3	50.8	47.5	51.2	51.7	50.1

Table 2.3. *H height in cm, W weight in kg*

EXERCISE 2.24 (MLE with $\Gamma(k, \lambda)$ distribution).– (see p. 187) Consider N observations X_n i.i.d., with:

1) distribution $\Gamma(1, \lambda)$ of density $p_X(x; \lambda) = \lambda^{-1} e^{-x/\lambda} \mathbb{1}(x \geq 0)$. Determine the expression of the MLE of λ. Perform a simulation. Compare the dispersions obtained to the Cramer-Rao bound (expression [2.108]);

2) distribution $\Gamma(k, \lambda)$; the expression of the distribution is given in [2.99]. Determine the expression of the MLE of $\theta = (k, \lambda)$.

EXERCISE 2.25 (Singularity in the MLE approach).– (see p. 188) Consider N observations i.i.d., with values in $\mathbb{R}$, with the probability distribution:

$$p_{X_n}(x_n; \theta) = \frac{1}{2\sqrt{2\pi}\sigma_0} e^{-(x_n - m_0)^2 / 2\sigma_0^2} + \frac{1}{2\sqrt{2\pi}\sigma_1} e^{-(x_n - m_1)^2 / 2\sigma_1^2}$$

where $\theta = (m_0, m_1, \sigma_0, \sigma_1) \in \Theta = \mathbb{R} \times \mathbb{R} \times \mathbb{R}^+ \times \mathbb{R}^+$. Show that the likelihood $\ell(\theta)$ may tend toward infinity. More precisely, for any $A > 0$, there exists $\theta \in \Theta$, such that $\ell(\theta) \geq A$.

EXERCISE 2.26 (Parameters of a homogeneous Markov chain).– (see p. 189) Consider a series of N Markovian r.v. X_0, ..., X_{N-1} with values in the finite set $\mathcal{S} = \{0, \ldots, S - 1\}$, with an initial distribution $\mathbb{P}\{X_0 = s\} = \alpha_s \geq 0$ and a transition law:

$$\mathbb{P}\{X_n = s | X_{n-1} = s'\} = p_{s|s'} \geq 0$$

Hence, $\sum_{s=0}^{S-1} \alpha_s = 1$ and for any $s' \in \{0, \ldots, S-1\}$, $\sum_{s=0}^{S-1} p_{s|s'} = 1$. For ease of calculation, the following notation may be used:

$$\mathbb{P}\{X_0 = x_0\} = \sum_{s=0}^{S-1} \alpha_s \mathbb{1}(x_0 = s)$$

and for $n = 1, \ldots, N - 1$:

$$\mathbb{P}\{X_n = x_n | X_{n-1} = x_{n-1}\} = \sum_{s=0}^{S-1} \sum_{s'=0}^{S-1} p_{s|s'} \mathbb{1}(x_n = s, x_{n-1} = s')$$

where x_n belong to $\mathcal{S}$.

Let us consider the function $g(x) = \sum_{s=0}^{S-1} g_s \mathbb{1}(s=x)$. Then, for any function f:

$$f(g(x)) = \sum_{s=0}^{S-1} f(g_s)\mathbb{1}(x=s)$$

1) Determine the expression of the likelihood associated with the observations X_0, ..., X_{N-1} as a function of $p_{s|s'}$ and α_s.

2) Use this result to deduce a maximum likelihood estimator for $p_{s|s'}$.

3) Write a program to generate a sequence of N Markov r.v.s with values in $\{0, \ldots, S-1\}$ for a transition law P (drawn at random) and for an initial distribution α (drawn at random), and then estimate the matrix P.

2.6.5.3. *EM algorithm*

Consider an observation Y with likelihood function $p_Y(y; \theta)$, the maximization of which is intractable. In contrast, we assume that a joint probability law $p_{X,Y}(x, y; \theta)$ exists, such that:

$$\mathcal{L}(\theta) = p_Y(y; \theta) = \int_{\mathcal{X}} p_{X,Y}(x, y; \theta) dx$$

In this context, the pair (X, Y) is known as *complete data* and X as *incomplete data*. Because $\log p_{X,Y}(x, y; \theta)$ cannot be maximized directly, since X has not been observed, the idea consists of maximizing the conditional expectation $\mathbb{E}\{\log p(X, Y; \theta)|Y\}$, which is, by definition, a function of Y and is therefore observable. The EM (expectation-maximization) algorithm that is used to solve this problem has been presented in a seminal article [DEM 77]. Each iteration involves the following two steps:

Data: $p_Y(y; \theta)$, $p_{X,Y}(x, y; \theta)$ distributions
Result: $\widehat{\theta}$
Initialization: $p = 0$, $\widehat{\theta}^{(0)}$;
while *stopping condition* **do**
    Expectation step:
        $Q(\theta, \theta^{(p)}) = \mathbb{E}_{\theta^{(p)}}\{\log p_{X,Y}(X, Y; \theta)|Y\}$;
    Maximization step:
        $\theta^{(p+1)} = \arg\max_\theta Q(\theta, \theta^{(p)})$;
    $p = p + 1$;
end

Algorithm 5: EM algorithm

$\theta^{(p)}$ denotes the estimated value of the parameter at the pth iteration of the algorithm. This two-step process is repeated until convergence is reached. Because the EM algorithm is only able to reach a local maximum, the choice of the initial value of θ is crucial. It is commonly advised to choose random starting points and keep the value that gives the highest maximized likelihood.

The following property is fundamental:

PROPERTY 2.4.– For each iteration of the EM algorithm, the likelihood of the observations increases. This can be written as:

$$p_Y(Y; \theta^{(p+1)}) \geq p_Y(Y; \theta^{(p)}) \qquad [2.121]$$

PROOF.– Indeed, if θ is such that $Q(\theta, \theta^{(p)}) \geq Q(\theta^{(p)}, \theta^{(p)})$, then using Bayes' rule, $p_{X,Y}(X, Y; \theta) = p_{X|Y}(X, Y; \theta) p_Y(y; \theta)$, we have:

$$\begin{aligned}
0 \leq Q(\theta, \theta^{(p)}) &- Q(\theta^{(p)}, \theta^{(p)}) \qquad [2.122] \\
&= \mathbb{E}_{\theta^{(p)}} \left\{ \log p_{X|Y}(X, Y; \theta) | Y \right\} + \log p_Y(Y; \theta) \\
&\quad - \mathbb{E}_{\theta^{(p)}} \left\{ \log p_{X|Y}(X, Y; \theta^{(p)}) | Y \right\} - \log p_Y(Y; \theta^{(p)}) \\
&= \mathbb{E}_{\theta^{(p)}} \left\{ \log \frac{p_{X|Y}(X, Y; \theta)}{p_{X|Y}(X, Y; \theta^{(p)})} \bigg| Y \right\} + \log p_Y(Y; \theta) - \log p_Y(Y; \theta^{(p)})
\end{aligned}$$

Using the concavity of the log function and the Jensen inequality, we obtain:

$$\mathbb{E}_{\theta^{(p)}} \left\{ \log \frac{p_{X|Y}(X, Y; \theta)}{p_{X|Y}(X, Y; \theta^{(p)})} \bigg| Y \right\} \leq \log \mathbb{E}_{\theta^{(p)}} \left\{ \frac{p_{X|Y}(X, Y; \theta)}{p_{X|Y}(X, Y; \theta^{(p)})} \bigg| Y \right\} = 0$$

Substituting this result in [2.122], we obtain:

$$\log p_Y(Y; \theta^{(p)}) - \log p_Y(Y; \theta) \geq Q(\theta, \theta^{(p)}) - Q(\theta^{(p)}, \theta^{(p)})$$

Therefore, choosing θ such that $Q(\theta, \theta^{(p)}) - Q(\theta^{(p)}, \theta^{(p)}) \geq 0$ leads to expression [2.121]. ∎

Based on the result, we also see that the likelihood of the observations increases as long as $Q(\theta, \theta^{(p)})$ increases. Therefore it is not necessary to take the maximum of $Q(\theta, \theta^{(p)})$ w.r.t. θ, as that is done in algorithm 5, but $Q(\theta, \theta^{(p)})$ can simply be increased. In this case, the algorithm is known as the generalized expectation maximization (GEM) algorithm.

Two examples of the applications of the EM algorithm are shown below, namely the mixture model and the censored data model.

2.6.5.4. *Mixture model*

Consider a series of N observations i.i.d., with a probability density written as:

$$p_{Y_n}(y_n; \theta) = \sum_{k=0}^{K-1} \alpha_k f_k(y_n; \mu_k) \qquad [2.123]$$

where, for example,

$$f_k(y_n; \mu_k) = \frac{1}{\sigma_k \sqrt{2\pi}} e^{-(y_n - m_k)^2 / 2\sigma_k^2}, \quad \text{where } \mu_k = (m_k, \sigma_k) \qquad [2.124]$$

The parameter of interest is therefore $\theta = \{m_0, \sigma_0, \alpha_0, \ldots, m_{K-1}, \sigma_{K-1}, \alpha_{K-1}\}$, where $\alpha_k \geq 0$, $\sum_k \alpha_k = 1$ and $\sigma_k \in \mathbb{R}^+$. This model is known as the Gaussian mixture model (GMM). It is used in a variety of fields, particularly speaker recognition and population mixing. A distribution of this type is shown in Figure 2.10. There are three discernible "states" or "modes": one around a value of 1, the second around a value of 5 and the third around a value of 8. In certain cases, there is a clear explanation for the presence of these "modes". For example, using a problem concerning the size of adult individuals, if two modes occur, these will clearly correspond to the male and female populations.

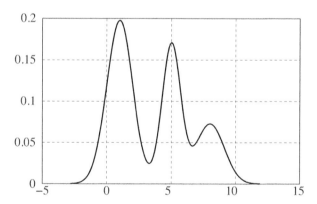

Figure 2.10. *Multimodal distribution*

From [2.123], the analytical expression of maximum likelihood is given by:

$$\widehat{\theta} = \arg \max_{\theta} \sum_{n=0}^{N-1} \log \sum_{k=0}^{K-1} \alpha_k f_k(y_n; \mu_k) \qquad [2.125]$$

Unfortunately, in this case, the calculation is intractable. This leads us to use the EM algorithm, which consists of an iterative search process for a local maximum of the likelihood function $\log p_{Y_1,\ldots,Y_{N-1}}(y_0,\ldots,y_{N-1};\theta)$.

Consider a series of random variables S_n i.i.d., with values in $\{0,\ldots,K-1\}$ and such that $\mathbb{P}\{S_n = k\} = \alpha_k$. We note:

$$p_{Y_n,S_n}(y_n, S_n = s_n; \theta) = \sum_{k=0}^{K-1} \alpha_k f_k(y_n; \mu_k) \mathbb{1}(s_n = k) \qquad [2.126]$$

We then verify that $\sum_{s_n=0}^{K-1} p_{Y_n,S_n}(y_n, S_n = s_n; \theta) = \sum_{k=0}^{K-1} \alpha_k f_k(y_n; \mu_k)$, which is expression [2.123] of the mixture model. We may therefore consider the pair (S_n, Y_n) as complete data. Taking the logarithm of the complete law [2.126], we obtain:

$$\ell(s, y; \theta) = \sum_{n=0}^{N-1} \sum_{k=0}^{K-1} \log(\alpha_k f_k(y_n; \mu_k)) \mathbb{1}(s_n = k)$$

$$= \sum_{n=0}^{N-1} \sum_{k=0}^{K-1} \log(f_k(y_n; \mu_k)) \mathbb{1}(s_n = k) + \sum_{n=0}^{N-1} \sum_{k=0}^{K-1} \log(\alpha_j) \mathbb{1}(s_n = k)$$

We will now determine the expression of the auxiliary function Q. Note that as the pairs (S_n, Y_n) are independent, the conditional expectation of S_n conditionally to $Y_{0:N-1}$ is only dependent on Y_n. Using the expression of f_k given by [2.124] and letting $v_k = \sigma_k^2$, we obtain:

$$Q(\theta, \theta^{(p)}) = \mathbb{E}_\theta\{\ell(S, Y; \theta)|Y\}$$

$$= -\sum_{n=0}^{N-1} \sum_{k=0}^{K-1} \frac{1}{2}\left(\log(v_k) + \frac{(y_n - m_k)^2}{v_k}\right) \mathbb{E}_{\theta^{(p)}}\{\mathbb{1}(S_n = k)|Y_n\}$$

$$+ \sum_{n=0}^{N-1} \sum_{k=0}^{K-1} \log(\alpha_k) \mathbb{E}_{\theta^{(p)}}\{\mathbb{1}(S_n = k)|Y_n\} - \frac{1}{2}NK\log(2\pi)$$

The expression of $\mathbb{E}_{\theta^{(p)}}\{\mathbb{1}(S_n = k)|Y_n\}$ is obtained using Bayes' rule:

$$\mathbb{E}_{\theta^{(p)}}\{\mathbb{1}(S_n = k)|Y_n\} = \mathbb{P}_{\theta^{(p)}}\{S_n = k|Y_n\}$$

$$= \frac{p_{Y_n|S_n=k}(y_n; \theta^{(p)})\mathbb{P}\{S_n = k; \theta^{(p)}\}}{p_{Y_n}(y_n; \theta^{(p)})}$$

$$= \frac{\alpha_k^{(p)} f_k(y_n; \mu_k^{(p)})}{\sum_{j=0}^{K-1} \alpha_j^{(p)} f_j(y_n; \mu_j^{(p)})}$$

We must therefore simply calculate $\alpha_k^{(p)} f_k(y_n; \mu_k^{(p)})$ and then normalize using:

$$c_n^{(p)} = p_{Y_n}(y_n; \theta^{(p)}) = \sum_{j=0}^{K-1} \alpha_j^{(p)} f_j(y_n; \mu_j^{(p)}) \qquad [2.127]$$

In what follows, we will use:

$$\gamma_{k,n}^{(p)} = \mathbb{P}_{\theta^{(p)}} \{S_n = k | Y_n\} \qquad [2.128]$$

which verifies that, for any n, $\sum_{k=0}^{K-1} \gamma_{k,n}^{(p)} = 1$ and thus $\sum_{n=0}^{N-1} \sum_{k=0}^{K-1} \gamma_{k,n}^{(p)} = N$.

We can now determine the value of the parameter that maximizes the function $Q(\theta, \theta^{(p)})$, canceling the derivatives with respect to θ.

For a maximization with respect to α, we consider the Lagrange multiplier η, associated with the constraint $\sum_k \alpha_k = 1$. Canceling the derivative with respect to α_k of the Lagrangian:

$$\mathcal{L} = Q(\theta, \theta^{(p)}) + \eta \left(1 - \sum_{j=0}^{K-1} \alpha_j\right)$$

we obtain:

$$\frac{1}{\alpha_k} \sum_{n=0}^{N-1} \mathbb{E}_{\theta^{(p)}} \{\mathbb{1}(S_n = k) | Y_n\} = \eta$$

hence, after normalization:

$$\alpha_k^{(p+1)} = \frac{1}{N} \sum_{n=0}^{N-1} \gamma_{n,k}^{(p)}$$

Substituting this value into function Q, we obtain the function $\widetilde{Q}$. Canceling the derivative of $\widetilde{Q}$ with respect to m_k, we have:

$$m_k^{(p+1)} = \frac{\sum_{n=0}^{N-1} \gamma_{k,n}^{(p)} y_n}{\sum_{n=0}^{N-1} \gamma_{k,n}^{(p)}}$$

Then, canceling the derivative with respect to v_k, we obtain:

$$v_k^{(p+1)} = \frac{\sum_{n=0}^{N-1} \gamma_{k,n}^{(p)} \left(y_n - m_k^{(p+1)}\right)^2}{\sum_{n=0}^{N-1} \gamma_{k,n}^{(p)}}$$

Finally, note that following expression [2.127], the log-likelihood of the observations for each iteration is written as:

$$\ell^{(p)} = \sum_{n=0}^{N-1} \log p_{Y_n}(y_n; \theta) = \sum_{n=0}^{N-1} c_n^{(p)} \qquad [2.129]$$

It is worth noting that $\ell^{(p)}$ must increase at each iteration of the EM algorithm.

The previous algorithm may be easily generalized to the cases with a mixture of K Gaussians of dimension d, with respective means m_0, ..., m_{K-1} and covariances C_0, ..., C_K. We note:

$$p(X; m, C) = \frac{1}{(2\pi)^{d/2}\sqrt{\det\{C\}}} e^{-\frac{1}{2}(X-m)^T C^{-1}(X-m)}$$

The algorithm is written as:

Data: K and $X_n \in \mathbb{R}^d$ for $n = 0$ to $N-1$
Result: $\widehat{\alpha}, \widehat{m}, \widehat{C}, \ell$
Initialization: $p = 0$, $\alpha_{0:K-1}^{(0)}, m_{0:K-1}^{(0)}, C_{0:K-1}^{(0)}$;
while *stopping condition* **do**
  for $n = 0$ *to* $N - 1$ **do**
    for $k = 0$ *to* $K - 1$ **do**
      $\quad g_{k,n} = p(X_n; m_{0:K-1}^{(p)}, C_{0:K-1}^{(p)})$;
    end
    $c_n = \sum_{k=0}^{K-1} \alpha_k^{(p)} g_{k,n}$;
    for $k = 0$ *to* $K - 1$ **do**
      $\quad \gamma_{k,n} = \alpha_k^{(p)} g_{k,n} / c_n$;
    end
  end
  $\alpha_k^{(p+1)} = \sum_{n=0}^{N-1} \gamma_{k,n} / N$;
  $\ell^{(p+1)} = \sum_{n=0}^{N-1} c_n$;
  for $k = 0$ *to* $K - 1$ **do**
    $S_k = \sum_{n=0}^{N-1} \gamma_{k,n}$
    $m_k^{(p+1)} = S_k^{-1} \sum_{n=0}^{N-1} \gamma_{k,n} X_n$
    $C_k^{(p+1)} = S_k^{-1} \sum_{n=0}^{N-1} \gamma_{k,n} (X_n - m_k^{(p+1)})(X_n - m_k^{(p+1)})^T$
  end
  $p = p + 1$;
end

Algorithm 6: EM algorithm for GMM

REMARKS:

– initialization may be carried out either at random or using a rough estimation, for example obtained by dividing the data into K blocks and calculating the empirical means and covariances for each block;

– the stopping condition usually involves the relative increase of $\ell^{(p)}$ as:

$$\rho = \frac{|\ell^{(p+1)} - \ell^{(p)}|}{|\ell^{(p)}|}$$

and/or a maximum number of iterations;

EXERCISE 2.27 (GMM generation).– (see p. 190) Write a program to:

– generate N data of a mixture of K Gaussians with respective proportions α_k, means m_k and variances σ_k^2;

– estimate the parameters of a GMM using the EM algorithm. For the initialization, divide the observations into K groups of the same size and estimate the mean and the variance for each group. For the proportion initialization, take $1/K$ for the K groups.

EXERCISE 2.28 (Estimation of states of a GMM).– (see p. 193)

Consider the mixture model used in exercise 2.27. We presume that the parameters of the model have been estimated, for example, using the program written in exercise 2.27.

– Using equation [2.128], propose an estimator of state S;

– Write a program to estimate the state of a series of observations and compare this state with the true value of the data obtained. Begin by applying a learning process for θ using a sample of 10000 values.

2.6.5.5. *Censored data model*

In this section, we will consider the distribution of survival periods of patients receiving treatment. The survival period for each patient is collected at a given moment. There are two possibilities: either the patient is deceased and the survival period is therefore known, or the patient is still living, in which case we know only that the survival period is greater than the observed treatment time. Observations of this type are said to be *right-censored*.

More precisely, let us denote X_n as the r.v. modeling the survival period and $p_X(x;\theta)$ as its probability density. We observe a sequence of N pairs (y_n, c_n), such that when $c_n = 0$, the value y_n is a realization of the r.v. X_n and when $c_n = 1$, we only know that the non-observed r.v. X_n is greater than y_n; in other words, $X_n \in (y_n, +\infty)$. Note that when $c_n = 0$, the likelihood of X_n is equal to $p_X(y_n; \theta)$, whereas when $c_n = 1$, this is no longer possible, and we only know that

$X_n \in (y_n, +\infty)$ with the probability of $(1 - F_X(y_n; \theta))$, where the cumulative function $F_X(x; \theta) = \int_{-\infty}^{x} p_X(u; \theta) du$.

An approach similar to the maximum likelihood method consists of maximizing the following expression:

$$\mathcal{L}(\theta) = \sum_{n=0}^{N-1} \log p_X(y_n; \theta) \mathbb{1}(c_n = 0) + \sum_{n=0}^{N-1} \log(1 - F_X(y_n; \theta)) \mathbb{1}(c_n = 1) \quad [2.130]$$

Generally speaking, it is impossible to maximize expression [2.130]; this leads us to use the EM algorithm. To do this, we consider that the full model is the survival value X_n, and thus:

$$Q(\theta, \theta') = \mathbb{E}_{\theta'} \{\log p_X(X_0, \ldots, X_N - 1; \theta) | Y_0, \ldots, Y_{N-1}, c_0, \ldots, c_{N-1}\}$$

where X_n are complete data points and Y_n are incomplete data points. Using the independence hypothesis, we obtain:

$$Q(\theta, \theta') = \sum_{n=0}^{N-1} \mathbb{E}_{\theta'} \{\log p_{X_n}(X_n; \theta) | Y_n, c_n\}$$

– when $c_n = 0$, as $X_n = y_n$, the conditional expectation of $\log p_X(X_n; \theta)$ conditionally on Y_n, based on the definition of conditional expectation itself, is equal to $\log p_X(y_n; \theta)$;

– when $c_n = 1$, the conditional expectation of $p_X(X_n; \theta)$ conditionally on the fact that $X_n > y_n$ is written as:

$$\mathbb{E}_{\theta'} \{\log p_X(X_n; \theta) | X_n > y_n\} = g(y_n)$$

where

$$g(y) = \frac{\int_{y}^{+\infty} \log p_X(x; \theta) p_X(x; \theta') dx}{1 - F_X(y; \theta')} \quad [2.131]$$

Applying Bayes' rule, we obtain:

$$\mathbb{P}_{\theta'} \{X_n \leq x | X_n > y\} = \frac{\mathbb{P}_{\theta'} \{X_n \leq x, X_n > y\}}{\mathbb{P}_{\theta'} \{X_n > y\}}$$

$$= \frac{\mathbb{P}_{\theta'} \{y < X_n \leq x\}}{\mathbb{P}_{\theta'} \{X_n > y\}} \mathbb{1}(y < x)$$

$$= \frac{\int_{y}^{x} p_X(u; \theta') du}{1 - F_X(y; \theta')} \mathbb{1}(y < x)$$

Consequently, applying a derivation with respect to x, we obtain the density of X_n conditionally on $X_n > y_n$, giving:

$$p_{X_n|X_n>y_n}(x,y;\theta') = \frac{p_X(x;\theta')}{1-F_X(y;\theta')}\mathbb{1}(x>y)$$

Finally, using the fact that for any function $h(x)$ independent of y:

$$\mathbb{E}_{\theta'}\{h(X)|X \geq y\} = \int_{-\infty}^{+\infty} h(x)p_{X|X>y}(x;\theta')dx$$

Using $h(x) = \log p_X(x;\theta)$ (note that we use θ and not θ'), we obtain the result given in [2.131]. Finally, we have:

$$Q(\theta,\theta') = \sum_{n=0}^{N-1} \log p_{X_n}(y_n;\theta)\mathbb{1}(c_n=0) \qquad [2.132]$$
$$+ \sum_{n=0}^{N-1} \frac{\int_{y_n}^{+\infty} \log p_X(x;\theta)p_X(x;\theta')dx}{\int_{y_n}^{+\infty} p_X(x;\theta')dx}\mathbb{1}(c_n=1)$$

In cases where the cancellation of the derivative with respect to θ is hard to express, we may choose to use the GEM algorithm instead of the EM algorithm. This consists of calculating an increase in the function Q. This may be done using a number of steps from the gradient algorithm. The method of moments may be used for initialization (see section 2.6.4).

EXERCISE 2.29 (MLE on censored data).– (see p. 194)

This exercise considers an unusual case where the recurrence of the EM algorithm has an analytical solution. We suppose that, following medical treatment, patients' survival periods X_n follow an exponential distribution of parameter θ, i.e. $p_X(x;\theta) = \theta e^{-x\theta}\mathbb{1}(x \geq 0)$ and thus $\log p(x;\theta) = \log \theta - \theta x$. During the evaluation process, certain survival periods are censored as the patients are still living at the moment of evaluation.

1) Using expression [2.132], determine the function $Q(\theta, \theta')$;

2) Determine the recurrence over θ associated with the EM algorithm;

3) Perform a simulation to compare the estimator obtained in the previous question with an estimation leaving aside the indication of censored data and an estimation which only uses uncensored data, using results from exercise 2.24, question 1;

4) The data available in statsmodels.apidatasets.heart consist of the survival day number after receiving a heart transplant, the age of the patient and whether or not the survival time is censored. Without taking into account the age, estimate the life expectation.

EXERCISE 2.30 (Heart implant, model with exogenous variable).– (see p. 195) The data available in statsmodels.apidatasets.heart consist of the survival day number after receiving a heart transplant, the age of the patient and whether or not the survival time was censored. The survival number is considered as the endogenous variable and age as the exogenous variable. We consider the likelihood model:

$$\log p(x; \theta(a)) = \log \theta(t) - \theta(a)x$$

where $\theta(a) = \alpha_0 + \alpha_1 a$ does depend linearly on the age a. Using the expression [2.132] of the auxiliary function of the EM algorithm, determine the recursion on the parameters α_0 and α_1. Based on the EM algorithm, write a program to estimate α_0 and α_1 for the data.

2.6.6. Logistic regression

In this section, we consider that the explanatory variable falls within $\mathcal{X} = \mathbb{R}^p$ and that the qualitative response has values within $\mathcal{Y} = \{0, 1\}$. Logistic regression consists of modeling the probability of the response Y_n, conditionally on the explanatory variable X_n, as follows:

$$\mathbb{P}\{Y_n = y_n | X_n = x_n; \beta_0, \beta\} = \{\{2.2 \begin{array}{l} \dfrac{e^{\beta_0 + \beta^T x_n}}{1 + e^{\beta_0 + \beta^T x_n}} \text{ if } y_n = 0 \\ \dfrac{1}{1 + e^{\beta_0 + \beta^T x_n}} \text{ if } y_n = 1 \end{array} \quad [2.133]$$

where $\beta_0 \in \mathbb{R}$ and $\beta \in \mathbb{R}^p$. We let $\alpha = \begin{bmatrix} \beta_0 & \beta \end{bmatrix}^T$.

Writing the log-likelihood of N observations, assumed to be independent, we obtain $p_{X,Y}(x, y; \alpha) = \prod_{n=0}^{N-1} p(y_n | x_n; \alpha) p_{X_n}(x_n)$. Assuming that the non-specified marginal distribution of X does not depend on α, the log-likelihood may be written, up to an additive constant that does not depend on α:

$$\ell(\alpha) = \sum_{n=0}^{N-1} \alpha^T Z_n \mathbb{1}(Y_n = 0) - \sum_{n=0}^{N-1} \log\left(1 + e^{\alpha^T Z_n}\right) \quad [2.134]$$

where $Z_n = \begin{bmatrix} 1 & X_n^T \end{bmatrix}^T$.

The search for a maximum likelihood estimator is a nonlinear maximization problem. It may be numerically solved, for example using the Newton-Raphson method.

The Newton-Raphson algorithm is an iterative algorithm that aims to minimize a real-valued function $\ell(\alpha)$ with respect to the multivariate α. Let α_p be the value calculated in step p. Thus, the value at step $p+1$ can be written as:

$$\alpha_{p+1} = \alpha_p - \left[\frac{\partial^2 \ell}{\partial^2 \alpha}\right]^{-1}_{\alpha=\alpha_p} \times \left.\frac{\partial \ell}{\partial \alpha}\right|_{\alpha=\alpha_p} \quad [2.135]$$

When maximizing expression [2.134], the first and second derivatives are written as:

$$\frac{\partial \ell}{\partial \alpha} = \sum_{n=0}^{N-1} Z_n \mathbb{1}(Y_n = 0) - \sum_{n=0}^{N-1} Z_n \frac{e^{\alpha^T Z_n}}{1 + e^{\alpha^T Z_n}}$$

$$= -\sum_{n=0}^{N-1} Z_n \mathbb{1}(Y_n = 1) + \sum_{n=0}^{N-1} \frac{1}{1 + e^{\alpha^T Z_n}} \quad [2.136]$$

and

$$\frac{\partial^2 \ell}{\partial^2 \alpha} = -\sum_{n=0}^{N-1} Z_n Z_n^T \frac{e^{\alpha^T Z_n}}{(1 + e^{\alpha^T Z_n})^2} \quad [2.137]$$

This algorithm is implemented in exercise 2.31.

While the logistic model is not identically distributed, property [2.107] may still be used, where

$$F(\alpha) = \lim_{N \to +\infty} \frac{1}{N} \sum_{n=0}^{N-1} Z_n Z_n^T \frac{e^{\alpha^T Z_n}}{(1 + e^{\alpha^T Z_n})^2}$$

$$\approx \frac{1}{N} \sum_{n=0}^{N-1} Z_n Z_n^T \frac{e^{\alpha^T Z_n}}{(1 + e^{\alpha^T Z_n})^2}$$

$\widetilde{Z}$ is the matrix of dimension $N \times (p+1)$, for which line n has the following expression:

$$\widetilde{Z}_n = \frac{e^{Z_n^T \alpha/2}}{1 + e^{Z_n^T \alpha}} Z_n$$

Thus, $F(\alpha) \approx N^{-1}\widetilde{Z}^T\widetilde{Z}$. From this, we see that:

$$(\widehat{\alpha}_{\mathrm{MLE}} - \alpha) \xrightarrow{d} \mathcal{N}(0, (\widetilde{Z}^T\widetilde{Z})^{-1}) \qquad [2.138]$$

This expression allows the calculation of asymptotic confidence intervals for the components of α, along with test statistics following the construction presented in the subsection on page 54. These results are applied in exercise 2.31.

EXERCISE 2.31 (Logistic regression).– (see p. 196) Write a program to estimate the parameter α of the logistic model, based on the iteration [2.135] of the Newton-Raphson algorithm.

Apply this program to the data series presented in Table 2.4, which gives the state, faulty/no faulty, of an O-ring (toric joint) as a function of temperature. Using [2.138], calculate a confidence interval at 95% of α. Using [2.30], calculate the p-value of the log-GLRT of the hypothesis $H_0 = \{\alpha_2 = 0\}$.

K	53	56	57	63	66	67	67	67	68	69	70	70
S	1	1	1	0	0	0	0	0	0	0	0	1
K	70	70	72	73	75	75	75	76	78	79	80	81
S	1	1	0	0	0	1	0	0	0	0	0	0

Table 2.4. *Temperature in degrees Fahrenheit and state of the O-ring: 1 signifies the existence of a fault and 0 the absence of faults. These data are related to the Space Shuttle Challenger disaster occurred on January 28, 1986*

EXERCISE 2.32 (GLRT for the logistic model).– (see p. 199) The function determined in exercise 2.31 calculates the log-likelihood of a logistic model based on a set of observations. It may therefore be used to determine the log-GLRT [2.29] of the hypothesis that one of the coefficients of model [2.133] is null.

Consider the logistic model where the parameter $\alpha = \begin{bmatrix} 0.3 & 0.5 & 1 & 0 & 0 \end{bmatrix}$. Let us note that the two last components of α are null. Write a program to test the hypothesis $H_0 = \{\alpha : \alpha_4 = \alpha_5 = 0\}$ with the GLRT approach and verify, in accordance with [2.30], that the statistic of the test has a χ^2 distribution with 2 degrees of freedom.

EXERCISE 2.33 (Logistic regression on home owner/rent).– (see p. 199)

Apply the function `logisticNR` proposed in exercise 2.31 to the data provided in `statsmodels.api.datasets.ccard`. The regression will be based on the following two explanatory variables: age and income.

2.6.7. *Non-parametric estimation of probability distribution*

This section provides a simplified discussion of non-parametric estimations of probability densities and cumulative functions. We will not consider kernel-based methods, which are essential when searching for a consistent estimation of probability density. The variance of an estimator may be reduced by carrying out smoothing using a kernel function.

2.6.7.1. *Histogram approach*

Consider a sample of N real-valued i.i.d. r.v.s $X_0, \ldots, X_{N-1}$, the probability density of which is denoted by $p(x)$. The approach considered here to estimate $p(x)$ consists of dividing the observation set into g intervals in a similar manner to constructing goodness-of-fit tests in section 2.5.3. Here the same notation will be used.

The total observation interval is split into g intervals Δ_j of respective length $\ell(\Delta_j)$. We denote c_j as the middle of Δ_j. Assuming that $p(x)$ is almost equal to $p(c_j)$ in all the intervals Δ_j, we can therefore write:

$$\mathbb{P}\{X_n \in \Delta_j\} = \int_{\Delta_j} p(x)dx \approx \ell(\Delta_j)\, p(c_j)$$

In contrast, based on the empirical distribution, we obtain an estimator of $\mathbb{P}\{X_n \in \Delta_j\}$, which is written as:

$$\widehat{P}_j = \frac{1}{N} \sum_{k=0}^{N-1} \mathbb{1}(X_k \in \Delta_j)$$

For a more detailed definition of the empirical distribution, see section 2.6.8, page 107.

Equalizing the last two expressions, we obtain an estimator for $p(c_j)$, written as:

$$\widehat{p}(c_j) = \frac{1}{\ell(\Delta_j)} \widehat{P}_j = \frac{\sum_{k=0}^{N-1} \mathbb{1}(X_k \in \Delta_j)}{N\ell(\Delta_j)}$$

Exercise 2.7 shows that the asymptotic distribution of a vector $\widehat{P}$, with g components $\widehat{P}_j$, converges in law toward a vector P, with g components of values $\mathbb{P}\{X_n \in \Delta_j\}$:

$$\sqrt{N}(\widehat{P} - P) \xrightarrow{d} \mathcal{N}(0, C)$$

where $C = \mathrm{diag}(P) - PP^T$. Hence:

$$\sqrt{N}(\widehat{p}(c_j) - p(c_j)) \xrightarrow{d} \mathcal{N}(0, \gamma_j)$$

where $\gamma_j = C_{jj}/\ell^2(\Delta_j)$. This expression allows us to calculate the confidence intervals of $p(c_j)$ of the form:

$$\widehat{p}(c_j) - \frac{c(\alpha)\sqrt{\gamma_j}}{\sqrt{N}} \leq p(c_j) \leq \widehat{p}(c_j) + \frac{c(\alpha)\sqrt{\gamma_j}}{\sqrt{N}}$$

where α is given by $\int_{-\infty}^{c} (2\pi)^{-1/2} e^{-u^2/2} du = 1 - \alpha/2$. In practice, γ_j is replaced by an estimate obtained from $\widehat{P}$.

The following program provides an example:

```
# -*- coding: utf-8 -*-
"""
Created on Thu Jun  2 07:27:34 2016
****** estimdsproba
@author: maurice
"""
from scipy.stats import norm
from numpy import linspace, sqrt
from numpy.random import randn
from matplotlib import pyplot as plt
N=10000;g=50; alpha=0.05; c=norm.isf(1.0-alpha/2.0);
x=linspace(-5.0,5.0,g); Deltax = x[1]-x[0]; V=randn(N); plt.clf()
auxH = plt.hist(V,x,normed='True'); hatP = auxH[0]; LP = len(hatP)
xP = auxH[1][0:LP]+(auxH[1][1]-auxH[1][0])/2.0
hatCP = hatP-hatP*hatP; CI95=c*sqrt(hatCP)/sqrt(N)/Deltax;
pfdOI=hatP-CI95; pfdOS=hatP+CI95; pdftheo=norm.pdf(xP,0,1);
plt.hold('on'); plt.plot(xP,pdftheo,'.-b');
plt.plot(xP,pfdOI,'.-r'); plt.plot(xP,pfdOS,'.-r');
plt.hold('off'); plt.grid('on')
```

2.6.7.2. *Estimation of the cumulative function*

Consider a sample of N i.i.d. r.v.s X_0, ..., X_{N-1}. We denote $F_X(x)$ as the common cumulative function, which is defined by $F(x) = \mathbb{P}\{X_n \leq x\}$. The empirical distribution method consists of assigning a probability of $1/N$ to each observed value. Therefore, an estimator of the cumulative function is expressed as:

$$\widehat{F}_N(x) = \frac{1}{N} \sum_{n=0}^{N-1} \mathbb{1}(X_n \leq x) \qquad [2.139]$$

Let us denote $\{X_{(n)}\}$ as the sequence of values sorted in the ascending order. The sequence $\{X_{(n)}\}$ is commonly called the *order statistic* of the sequence $\{X_n\}$. Using the *order statistic* of X_n, we can rewrite expression [2.139] as:

$$\widehat{F}_N(s) = \frac{1}{N}\mathbb{1}(X_{(0)} \leq s < X_{(1)}) + \frac{2}{N}\mathbb{1}(X_{(1)} \leq s < X_{(2)}) + \cdots \quad [2.140]$$

Therefore, $\widehat{F}_N(s)$ appears as a stepwise function, with steps equal to $1/N$ and located on the values of the sequence $\{X_{(n)}\}$. From this, we deduce the algorithm for performing $\widehat{F}_N(s)$: (i) ordering the values in ascending order and (ii) assigning a step $1/N$ to the ordered values.

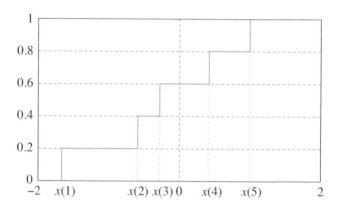

Figure 2.11. *Step equal to $1/N$ for the five values of the series, ranked in increasing order*

Following [2.139], $\widehat{F}_N(x)$ appears as the mean of the sequence of i.i.d. r.v.s $Y_i = \mathbb{1}(X_i \leq x)$ with same means $F(x)$ and same variances $F(x)(1 - F(x))$. Therefore, the central limit theorem is written as:

$$\sqrt{N}\left(\widehat{F}_N(x) - F(x)\right) \rightarrow N(0, F(x)(1 - F(x))) \quad [2.141]$$

Hence, an approximate confidence interval at 95% can be obtained using:

$$\widehat{F}_N(x) - \frac{1.96\sqrt{\widehat{\gamma}}}{\sqrt{N}} \leq F(x) \leq \widehat{F}_N(x) + \frac{1.96\sqrt{\widehat{\gamma}}}{\sqrt{N}}$$

where $\widehat{\gamma} = \widehat{F}_N(x)(1 - \widehat{F}_N(x))$.

It is worth noting that the empirical median can be derived from the order statistic.

DEFINITION 2.11 (Empirical median).– Let $\{X_{(0)}, \ldots, X_{(N-1)}\}$ be the *order statistic* of the sequence $\{X_0, \ldots, X_{N-1}\}$. The median can be written as:

$$M = \begin{cases} X_{((N-1)/2)} & \text{if } N \text{ is odd} \\ \frac{1}{2}(X_{(N/2-1)} + X_{(N/2)}) & \text{if } N \text{ is even} \end{cases}$$

EXERCISE 2.34 (Cumulative function estimation).– (see p. 200) We consider a sequence of i.i.d. r.v.s in the following two cases: (i) with discrete multinomial distribution with four respective probabilities 0.5, 0.25, 0.125 and 0.125 and (ii) with Gaussian continuous distribution with mean 0 and variance 1.

Write a program to generate a sequence of N values, estimate the cumulative function and compare the result to the theoretical result.

EXERCISE 2.35 (Estimation of a quantile).– (see p. 201) Consider an r.v. with a cumulative function denoted by $F(x)$. The quantile at $100c\%$ associated with $F(x)$ is defined by:

$$c \mapsto s = \min\{t : F(t) \geq c\} \qquad [2.142]$$

1) Use definition [1.5] to deduce an estimator $\widehat{s}_N$ of s;

2) Applying the δ-method (see section 1.5.3) to expression [2.141], deduce the asymptotic distribution of $\widehat{s}_N$;

3) Use this result to deduce an approximate confidence interval at 95%;

4) Write a program to verify these results.

Exercise 2.36 illustrates the estimation of the cumulative function in the context of image processing.

EXERCISE 2.36 (Image equalization).– (see p. 202) Image equalization in grayscale consists of transforming the value of pixels, so that the pixel value distribution of the transformed image is uniform. In section 4.3.1, we saw that the application of the function $F(x)$ to an r.v. of the cumulative function $F(x)$ gives an r.v. with a uniform distribution over $(0, 1)$. This result can be applied in equalizing images. To do this, we begin by estimating the cumulative function of the original image, and then applying the estimated distribution to the image.

Consider a grayscale image where the value $I_{n,m}$ of pixel (n, m) is modeled as an r.v. with values in the set $\mathcal{I} = \{0, \ldots, S-1\}$. Let p_s be the probability that a pixel will be equal to s, and $F(x)$ be the cumulative function. This is a piecewise linear function with steps located in $\mathcal{I}$ of height $p_s = \mathbb{P}\{I_{n,m} = s\}$ (see Figure 2.11). Thus, to estimate the cumulative function of the image, we estimate the probability p_s using the empirical law, for s from 0 to $S-1$, written as:

$$\widehat{p}_s = \frac{1}{NM} \sum_{n=0}^{N-1} \sum_{m=0}^{M-1} \mathbb{1}(I_{n,m} = s)$$

This gives us an estimate of the cumulative function:

$$\widehat{F}(x) = \sum_{s=0}^{S-1} \widehat{p}_s \mathbb{1}(s \leq x)$$

Write a program to equalize the levels of gray in an image. Verify the result by estimating the cumulative function of the transformed image. Note that the cumulative function of the uniform distribution over $(0,1)$ is written as $F(x) = x \times \mathbb{1}(x \in (0,1))$.

2.6.8. *Bootstrap and others*

In this section, we will give a brief introduction to the non-parametric estimation of estimator variance. In practice, three approaches are generally used: *bootstrap approach*, *jackknife approach* and *cross-validation*. In all the cases, once the variance has been estimated and based on the assumption of Gaussian behavior, it becomes possible to estimate a confidence interval.

Before considering the way in which estimator variance is determined, let us define the empirical distribution and expectation.

DEFINITION 2.12 (Empirical distribution).– Consider a sequence of N i.i.d. random vectors $X_0, \ldots, X_{N-1}$ taking their values in the space $\mathbb{R}^d$ (provided with its Borel set $\mathcal{B}^{\otimes d}$). The empirical distribution, derived from the observations $X_0, \ldots, X_{N-1}$, associates at each Borelian $b \in \mathcal{B}^{\otimes d}$ the probability

$$\widehat{\mathbb{P}}_N(b) = \sum_{n=0}^{N-1} \frac{1}{N} \mathbb{1}(X_n \in b) \qquad [2.143]$$

In short, each observation has an assigned probability of $1/N$. Moreover, taking into account that the X_n are i.i.d., we can express the joint distribution of any sequence of N Borelians $b_{n'} \in \mathcal{B}^{\otimes d}$ as the product of the identical empirical distributions, which can be written as:

$$\widehat{\mathbb{P}}_N^{(N)}(b_0, \ldots, b_{N-1}) = \prod_{n'=0}^{N-1} \sum_{n=0}^{N-1} \frac{1}{N^N} \mathbb{1}(X_n \in b_{n'})$$

This is equivalent to assigning a probability of $1/N^N$ to all of the N-uplets constructed, with duplication, using these observations.

It is worth noting that the empirical distribution is random due to the use of sequence X_n: each outcome ω is associated with a realization of the sequence X_0, ..., X_{N-1}, and thus with a realization of the empirical distribution.

Based on the joint empirical distribution, the empirical expectation of the function $g(X_0, \ldots, X_{N-1})$ with values in $\mathbb{R}^q$ is defined as follows.

DEFINITION 2.13 (Empirical expectation).– Consider a sequence of N i.i.d. random vectors $X_0, \ldots, X_{N-1}$ with their values in $\mathbb{R}^d$ given its Borel set $\mathcal{B}^{\otimes d}$. The empirical distribution associated with this sequence is denoted by $\widehat{\mathbb{P}}_N^{(N)}$. The empirical expectation of the function $g(X_0, \ldots, X_{N-1})$, with their values in $\mathbb{R}^q$, is the vector:

$$\widehat{G}_N = \mathbb{E}_{\widehat{\mathbb{P}}_N^{(N)}} \{g(X_0, \ldots, X_{N-1})\}$$

Using [2.144], it is easy to show that:

$$\mathbb{E}_{\widehat{\mathbb{P}}_N^{(N)}} \{g(X_0, \ldots, X_{N-1})\} = \frac{1}{N^N} \sum_{1 \leq i_0, \ldots, i_{N-1} \leq N} g(X_{i_0}, \ldots, X_{i_{N-1}}) \quad [2.144]$$

In the specific case where $g(X_1, \ldots, X_N) = \prod_{n=0}^{N-1} g_n(X_n)$, we have:

$$\widehat{G}_N = \frac{1}{N^N} \prod_{n=0}^{N-1} \sum_{n=0}^{N-1} g_n(X_n)$$

In the specific case where $g(X_0, \ldots, X_{N-1}) = g(X_0)$, we have:

$$\widehat{G}_N = \frac{1}{N} \sum_{n=0}^{N-1} g(X_n)$$

Now, consider a series of N observations i.i.d., of unknown distribution, and let θ be the parameter of interest and $\widehat{\theta}(X_0, \ldots, X_{N-1})$ an estimator of θ. To estimate the variance of $\widehat{\theta}$, expression [2.144] can be applied to calculate the first two moments of $\widehat{\theta}(X_0, \ldots, X_{N-1})$; however, this calculation is often impossible, even for simple estimators. Note that, in general, N^N terms are involved (a very large number of terms for $N = 100$, for example). This leads us to consider a Monte Carlo-type approach, which consists of carrying out a draw from these N^N possible cases; the bootstrap approach, introduced by B. Efron [EFR 79], may be seen in this way.

2.6.8.1. Bootstrap

Consider a sequence of observations $X_0, \ldots, X_{N-1}$ and a parameter of interest θ of dimension p. The bootstrap technique consists of carrying out B random draws, with replacement of N samples from the set of observations. B is always very small compared to N^N, typically $B = 300$. The samples of the bth drawing are denoted by $X_0^{(b)}, \ldots, X_{N-1}^{(b)}$, and the associated estimation value is denoted by $\widehat{\theta}(X_0^{(b)}, \ldots, X_{N-1}^{(b)})$. For the sake of simplicity, we note that $\widehat{\theta}(X^{(b)}) = \widehat{\theta}(X_0^{(b)}, \ldots, X_{N-1}^{(b)})$. The bootstrap technique may then be used to obtain the empirical covariance:

$$\widehat{\sigma}_B^2(\widehat{\theta}) = \frac{1}{B-1} \sum_{b=0}^{B-1} (\widehat{\theta}(X^{(b)}) - \widehat{\mu}_B(\widehat{\theta}))(\widehat{\theta}(X^{(b)}) - \widehat{\mu}_B(\widehat{\theta}))^T \qquad [2.145]$$

with $\quad \widehat{\mu}_B(\widehat{\theta}) = \dfrac{1}{B} \sum_{b=0}^{B-1} \widehat{\theta}(X^{(b)})$

Note that the term $(B-1)$ in [2.145] leads to the creation of an unbiased estimator of the covariance matrix.

EXAMPLE 2.11 (Bootstrap of mean estimator).– Consider a sequence of N Gaussian r.v.s of mean μ and variance 1, and consider the estimation of the mean $\widehat{\mu} = N^{-1} \sum_{n=0}^{N-1} X_n$. Write a program to compare the variance using the bootstrap technique and the theoretical variance of $\widehat{\mu}$, with a value of $1/N$, over L simulations.

HINTS: *Type:*

```
# -*- coding: utf-8 -*-
"""
Created on Wed Jun  1 08:02:15 2016
****** bootstraponmean
@author: maurice
"""
from numpy import zeros, mean, std
from numpy.random import randn, randint
from matplotlib import pyplot as plt
N = 300; mu = 3; B = 300; Lruns = 100;
sigma2b = zeros(Lruns); mub = zeros(B);
for irun in range(Lruns):
    X = randn(N)+mu;
    U = randint(0,N,size=[N,B]);
    Xb = X[U]; mub = mean(Xb,axis=0);
    sigma2b[irun]= std(mub)**2;
```

```
plt.clf()
plt.boxplot(sigma2b)
print(mean(sigma2b))
```

∎

EXERCISE 2.37 (Bootstrap for a regression model).– (see p. 203) Consider a series of N observations of the linear model:

$$X = Z\beta + \sigma\epsilon$$

where

$$Z = \begin{bmatrix} 1 & 0 \\ 1 & 1 \\ \vdots & \vdots \\ 1 & N-1 \end{bmatrix} \text{ and } \beta = \begin{bmatrix} 3 & 2 \end{bmatrix}^T$$

and ϵ is a centered Gaussian vector with covariance I_N. Note that, according to property 2.2, the covariance matrix of $\widehat{\beta} = (Z^T Z)^{-1} Z^T X$ is expressed as $(Z^T Z)^{-1}$. Write a program to compare the covariance obtained using the bootstrap technique and the theoretical covariance over L simulations.

2.6.8.2. *Jackknife*

The jackknife (multi-usage foldable knife) technique was introduced by M. H. Quenouille [QUE 56, MIL 74] to reduce the bias of an estimator. J. W. Tuckey [TUC 58] extended the technique for estimating the estimator variance.

The fundamental idea involves calculating $\widehat{\theta}$ several times, removing a sample each time. More precisely, if $X^{(j)}$ denotes the sequence of observations from which the sample X_j is removed, we obtain the following N estimates:

$$\widehat{\theta}^{(j)} = \widehat{\theta}(X^{(j)}) \qquad [2.146]$$

From this, we obtain the jackknife empirical covariance associated with the estimator $\widehat{\theta}$:

$$\widehat{\sigma}_J^2 = \frac{N-1}{N} \sum_{j=0}^{N-1} (\widehat{\theta}^{(j)} - \widehat{\mu}_J)(\widehat{\theta}^{(j)} - \widehat{\mu}_J)^T \qquad [2.147]$$

$$\text{with } \widehat{\mu}_J(\widehat{\theta}) = \frac{1}{N} \sum_{j=0}^{N-1} \widehat{\theta}^{(j)} \qquad [2.148]$$

A simple application is provided for the mean estimate:

$$\widehat{\theta}^{(j)} = \frac{1}{N-1} \sum_{i \neq j} X_i \quad [2.149]$$

2.6.8.3. Cross-validation

Cross-validation may be seen as a generalization of the jackknife approach. Instead of removing a single value (leave-one-out, LOO), we remove several values. Typically, the sequence of observations is divided into K blocks of the same length, and each of the K blocks is used successively as the test database, with the remaining $(K-1)$ blocks acting as the learning database. Let the block k be our test database. Over the remaining $(K-1)$ blocks, we estimate θ, obtaining the value of $\widehat{\theta}^{(-k)}$. We then calculate the error on the block k using the parameter value of $\widehat{\theta}^{(-k)}$. We then calculate an average over the K blocks (see Figure 2.12).

Without losing generality, we assume that $N = LK$, where L is the size of a block. The element of the block of test data k is denoted as $X_\ell^{(k)} = X_{(k-1)L+\ell}$, with ℓ ranging from 1 to L, and the remaining data is denoted as $X_\ell^{(-k)}$. The covariance of the estimator $\widehat{\theta}(X)$ is written as:

$$\widehat{\sigma}_{CV}^2 = \frac{1}{K} \sum_{k=1}^{K} \frac{1}{L-1} \sum_{\ell=1}^{L} (\widehat{\theta}(X_\ell^{(k)}) - \widehat{\theta}(X_\ell^{(-k)}))(\widehat{\theta}(X_\ell^{(k)}) - \widehat{\theta}(X_\ell^{(-k)}))^T \quad [2.150]$$

learning	learning	learning	testing	learning	learning

Figure 2.12. *Cross-validation: a block is selected as the test base, for example block no. 4 in this illustration, and the remaining blocks are used as a learning base, before switching over*

Instead of using the cross-validation scheme, as shown in Figure 2.12, another scheme consists of randomly drawing the learning subset and the testing subset. This can be done using the function sklearn.train_test_split, which splits arrays into random train and test subsets. The parameter test_size indicates the percentage of the full set used for the test subset. The balance is associated with the training subset.

The cross-validation approach may be used to estimate the order of the model. If we add columns to Z, the projection theorem guarantees that the average prediction error for the learning data can only decrease or, at worst, remain constant if the additional vector is contained in the space created by the columns of Z. In simple terms, the more the columns are added, the easier it is to "explain" the noise ϵ. However, the more the columns are added, the more the error over the test data will increase; this is illustrated in exercise 2.38.

EXERCISE 2.38 (Model selection based on cross-validation).– (see p. 203) We want to show that, if we increase the number of predictors, the error begins by decreasing and then increases. The minimum can be used to estimate the significative predictors. Write a program to:

– generate $N = 3000$ data following the linear model $y = Z\theta + \sigma\epsilon$, where Z is the design matrix associated with the first 10 columns of a matrix X of size $N \times 20$ obtained by randn(N,20). Therefore, the first 10 columns of X represent the explanatory variables, whereas the others are non-explanatory.

– apply a cross-validation of order $K = 10$ (see scheme 2.12) and vary the number of predictors from 1 to 20, calculate the prediction errors over the learning base and over the test base as a function of p.

EXERCISE 2.39 (Cross-validation on CO_2 concentration).– (see p. 204) Taking CO_2 concentration with the linear model considered in exercise 2.11, write a program to perform the residual standard deviation for different values of (p,q). Use this calculation to estimate p and q using a cross-validation approach: for each couple (p,q) in $[0,\ldots,6] \times [1,\ldots,6]$, the data will be split 100 times into 80% for training and 20% for testing.

EXERCISE 2.40 (Cross-validation for home owner/rent).– (see p. 206) Taking the data considered in exercise 2.33, write a program to estimate the logistic model from the learning database and derive the prediction on a testing database. The cross-validation approach will be obtained with the function sklearn.train_test_split using 80% for learning and 20% for testing, and will be run 500 times. At each time, the percentage of good prediction will be evaluated.

EXERCISE 2.41 (Model selection with cross-validation).– (see p. 206) Let us consider the file statsmodels.api.datasets.diabetes.load() of exercise 2.13. We let $y_n = \beta_0 + \sum_{k=1}^{10} \beta_k x_{n,k} + \sigma\epsilon_n$, where $x_{n,k}$ denotes the kth features associated with the patient n. Write a program to perform the backward stepwise algorithm 4 and select the model using the residual norms derived from the cross-validation scheme shown in Figure 2.12. Compare the result with the results of exercises 2.15 and 2.16.

3

Inferences on HMM

In this chapter, we will give a brief overview of hidden Markov models (HMM) [BAS 93]. These models are widely used in many areas. They have a fundamental property that results in the existence of recursive algorithms, meaning that the number of operations and the size of the memory needed to calculate the required values do not increase with the number of samples. The best-known example of this is the Kalman filter [EMI 60].

Throughout this chapter, for simplicity, the notation $(n_1 : n_2)$ will be used to denote the sequence of integer values from n_1 to n_2 inclusive.

3.1. Hidden Markov models (HMM)

A hidden Markov model (HMM) is a bivariate discrete-time process (X_n, Y_n), where X_n and Y_n are two real random vectors of finite dimension, such that:

– X_n, $n \geq 0$, is a Markov process, i.e. for any function f, the conditional expectation of $f(X_{n+1})$ given the σ-algebra generated by $\{X_s; s \leq n\}$ (the past until n) coincides with the conditional expectation of $f(X_{n+1})$ given the σ-algebra generated by $\{X_n\}$. If the conditional distributions have a density, it can be written as:

$$p_{X_{n+1}|X_{0:n}}(x_{n+1}; x_{0:n}) = p_{X_{n+1}|X_n}(x_{n+1}; x_n) \qquad [3.1]$$

– Y_n, $n \geq 0$, is a process such that the conditional distribution of $Y_0, \ldots, Y_{n-1}$ given $X_0, \ldots, X_{n-1}$ is the product of the distributions of Y_k conditionally on X_k. If the conditional distributions have a density, it can be written as:

$$p_{Y_{0:n}|X_{0:n}}(y_{0:n}; x_{0:n}) = \prod_{k=0}^{n} p_{Y_k|X_k}(y_k; x_k) \qquad [3.2]$$

– the initial r.v. X_0 has a known probability law. If this initial distribution has a probability density, it will be denoted as $p_{X_0}(x_0)$.

The following expression of the joint distribution can be deduced from the previous assumptions:

$$p_{X_{0:n},Y_{0:n}}(x_{0:n}, y_{0:n}) = \qquad \qquad \qquad \qquad \qquad \qquad [3.3]$$
$$\prod_{k=0}^{n} p_{Y_k|X_k}(y_k; x_k) \prod_{k=1}^{n} p_{X_k|X_{k-1}}(x_k; x_{k-1}) p_{X_0}(x_0)$$

PROOF.– Using Bayes' rule and equation [3.2], we have:

$$p_{X_{0:n},Y_{0:n}}(x_{0:n}, y_{0:n}) = p_{Y_{0:n}|X_{0:n}}(y_{0:n}; x_{0:n}) p_{X_{0:n}}(x_{0:n})$$
$$= \prod_{k=0}^{n} p_{Y_k|X_k}(y_k; x_k) p_{X_{0:n}}(x_{0:n})$$

Again, using Bayes' rule and equation [3.1], we may write:

$$p_{X_{0:n}}(x_{0:n}) = p_{X_n|X_{0:n-1}}(x_n; x_{0:n-1}) p_{X_{0:n-1}}(x_{0:n-1})$$
$$= p_{X_n|X_{n-1}}(x_n; x_{n-1}) p_{X_{0:n-1}}(x_{0:n-1})$$

By repeating this process, we deduce that:

$$p_{X_{0:n}}(x_{0:n}) = \prod_{k=1}^{n} p_{X_k|X_{k-1}}(x_k; x_{k-1}) p_{X_0}(x_0)$$

which completes the proof of expression [3.3]. ∎

Expression [3.3] may be represented by the directed acyclic graph (DAG) shown in Figure 3.1, using the coding rule:

$$\mathbb{P}\{X_{0:n}, Y_{0:n}\} = \prod_{i \in \mathcal{V}} \mathbb{P}\{i|\text{parents}(i)\}$$

where $\mathcal{V}$ denotes the set of all nodes in the graph. Note that the DAG of an HMM is a tree. In more general cases, we speak of a dynamic Bayesian network.

In practice, the variables $Y_{0:n}$ represent observations and variables $X_{0:n}$ represent "hidden" variables. Our objective is thus to make inferences concerning the hidden variables based on the observations. In very general terms, we therefore need to calculate conditional distributions of the form $p_{X_{n_1:n_2}|Y_{m_1:m_2}}(x_{n_1:n_2}; y_{m_1:m_2})$. Thus, if we wish to "extract" all information concerning X_n based on the observations $Y_{0:n}$, we need to determine the distribution $p_{X_n|Y_{0:n}}(x_n; y_{0:n})$. Subsequently, any function of interest $f(X_n)$ may be calculated using the conditional expectation:

$$\mathbb{E}\{f(X_n)|Y_{0:n}\} = \int f(x) p_{X_n|Y_{0:n}}(x; Y_{0:n}) dx$$

which is a "measurable" function of the observations.

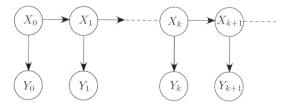

Figure 3.1. *Directed acyclic graph (DAG) associated with an HMM*

A priori, the problem appears very simple. We only need to apply Bayes' rule and write as follows:

$$p_{X_{n_1:n_2}|Y_{m_1:m_2}}(x_{n_1:n_2}; y_{m_1:m_2}) = \frac{p_{X_{n_1:n_2}, Y_{m_1:m_2}}(x_{n_1:n_2}, y_{m_1:m_2})}{p_{Y_{m_1:m_2}}(y_{m_1:m_2})}$$

Let us note that the numerator and the denominator can be obtained by integrating the joint probability distribution a certain number of times. For example, we have:

$$p_{X_n|Y_{0:n}}(x_n; y_{0:n}) = \frac{\int p_{X_{0:n}, Y_{0:n}}(x_{0:n}, y_{0:n}) dx_{0:n-1}}{\int p_{X_{0:n}, Y_{0:n}}(x_{0:n}, y_{0:n}) dx_{0:n}}$$

Unfortunately, with the exception of certain cases – such as the linear Gaussian case that leads to the Kalman filter, or the case where the variables $X_{0:n}$ have values within a discrete finite set – this expression is impossible to calculate. Other methods, such as extended Kalman filter, particles filtering, etc., are therefore required for these cases.

The factorized form of the expression [3.3] induces separation properties that are the basis of recursive algorithms. Let us consider a few examples. For any $k \leq n$, it is easy to show that:

$$p_{X_{0:n}, Y_{0:n}}(x_{0:n}, y_{0:n}) = \qquad [3.4]$$
$$p_{X_{0:k}, Y_{0:k}}(x_{0:k}, y_{0:k}) p_{X_{k+0:n}, Y_{k+0:n}|X_k}(x_{k+0:n}, y_{k+0:n}; x_k)$$

or that:

$$p_{X_{0:n}, Y_{0:n}}(x_{0:n}, y_{0:n}) = p_{X_{0:k}, Y_{0:k}}(x_{0:k}, y_{0:k}) \cdots$$
$$p_{X_{k+1}|X_k}(x_{k+1}; x_k) p_{X_{k+2:n}, Y_{k+2:n}|X_{k+1}}(x_{k+2:n}, x_{k+2:n}; x_{k+1})$$

or that, for any $j \geq 1$:

$$p_{X_{0:n}, Y_{0:n}}(x_{0:n}, y_{0:n})(x_{0:n}, y_{0:n}) = p(x_{0:k}, y_{0:k}) \cdots \qquad [3.5]$$
$$p_{X_{k+0:k+j}, Y_{k+0:k+j}|X_k}(x_{k+0:k+j}, y_{k+0:k+j}; x_k) \cdots$$
$$p_{X_{k+j+0:n}, Y_{k+j+0:n}|X_{k+j}}(x_{k+j+0:n}, y_{k+j+0:n}; x_{k+j})$$

Taking $j = 1$ and integrating expression [3.5] over $x_{0:k-1}$ and over $x_{k+2:n}$, we obtain:

$$p_{X_{k:k+1}, Y_{0:n}}(x_k, x_{k+1}, y_{0:n}) = p_{X_k|Y_{0:k}}(x_k; y_{0:k}) \qquad [3.6]$$
$$p_{Y_{k+1}|X_{k+1}}(y_{k+1}; x_{k+1}) p_{X_{k+1}|X_k}(x_{k+1}; x_k) p_{Y_{k+2:n}|X_{k+1}}(y_{k+2:n}; x_{k+1})$$

All of these results may be obtained using graphic rules applied to the DAG associated with the considered Bayesian network. For more details, refer to [PEA 88].

3.2. Inferences on HMM

The following list covers a number of inference problems:

– *Learning*: the joint law [3.3] is taken to depend on a parameter θ, which may be constituted of parameters μ associated with the distributions of observations $p_{Y_k|X_k}(y_k; x_k, \mu)$, and/or parameters ϕ associated with the distributions of the hidden states $p_{X_{k+1}|X_k}(x_{k+1}; x_k, \phi)$. The aim is to estimate θ based on a series of observations $Y_{0:n}$. The EM algorithm presented in section 2.6.5.3 may be used.

– *Inferences* on X_{n+k} based on the observations $Y_{0:n}$. Here the joint law [3.3] is assumed to be known, and we wish to find the expression of $p_{X_{n+k}|Y_{0:n}}(x_{n+k}; y_{0:n})$:

1) If $k = 0$, it is known as *filtering*;

2) If $k > 0$, it is known as *prediction*;

3) If $k < 0$, it is known as *smoothing*.

– *Estimation* of the sequence of hidden states $X_{0:n}$ based on the observations $Y_{0:n}$ in the case where the values of X_n are found in $\mathcal{S} = \{0, \ldots, S-1\}$, which is finite and discrete. To do this, the a posteriori maximum can be used:

$$\widehat{X}_{0:n} = \arg \max_{s_{0:n} \in \mathcal{S}^{n+1}} \mathbb{P}\{X_{0:n} = s_{0:n} | Y_{0:n}\}$$

One rough approach consists of "testing" the S^{n+1} possible value combinations. However, a deeper examination allows us to obtain an algorithm with a complexity in terms of only $(n+1) \times S^2$. This is the Viterbi algorithm, which is presented in section 3.5.5.

3.3. Filtering: general case

In general cases, filtering consists of calculating the *a posteriori* distribution $p_{X_n|Y_{0:n}}(x_n|y_{0:n})$. A simple calculation shows that the HMM structure leads to a recursive algorithm, which calculates $p_{X_{n+1}|Y_{0:n+1}}(x_{n+1}|y_{0:n+1})$ in two steps based on $p_{X_n|Y_{0:n}}(x_n|y_{0:n})$:

1) a *prediction step*, which calculates:

$$p_{X_{n+1}|Y_{0:n}}(x_{n+1};y_{0:n}) = \int p_{X_n|Y_{0:n}}(x_n;y_{0:n})p_{X_{n+1}|X_n}(x_{n+1};x_n)dx_n$$

2) an *update step*, which calculates:

$$p_{X_{n+1}|Y_{0:n+1}}(x_{n+1};y_{0:n+1})$$
$$= \frac{p_{X_{n+1}|Y_{0:n}}(x_{n+1};y_{0:n})p_{Y_{n+1}|X_{n+1}}(y_{n+1};x_{n+1})}{p_{Y_{n+1}|Y_{0:n}}(y_{n+1};y_{0:n})} \quad [3.7]$$

Note that the denominator $p_{Y_{n+1}|Y_{0:n}}(y_{n+1};y_{0:n})$ is not dependent on x_{n+1}. Hence, using the fact that the integral of $p_{X_{n+1}|Y_{0:n+1}}(x_{n+1};y_{0:n})$ with respect to x_{n+1} is equal to 1, we deduce that:

$$p_{Y_{n+1}|Y_{0:n}}(y_{n+1};y_{0:n})$$
$$= \int p_{X_{n+1}|Y_{0:n}}(x_{n+1};y_{0:n})p_{Y_{n+1}|X_{n+1}}(y_{n+1};x_{n+1})dx_{n+1}$$

Therefore, we only need to calculate the numerator of [3.7]:

$$p_{X_{n+1}|Y_{0:n}}(x_{n+1}; y_{0:n}) p_{Y_{n+1}|X_{n+1}}(y_{n+1}; x_{n+1})$$

and then normalize by integration with respect to x_{n+1}. Hence, we can replace, in the update step, expression [3.7] by:

$$p_{X_{n+1}|Y_{0:n+1}}(x_{n+1}; y_{0:n+1})$$
$$\propto p_{X_{n+1}|Y_{0:n}}(x_{n+1}; y_{0:n}) p_{Y_{n+1}|X_{n+1}}(y_{n+1}; x_{n+1}) \quad [3.8]$$

where the symbol $\propto$ means "proportional to" i.e. up to a function which does not depend on x_{n+1}. Usually this function can be obtained by using that the integral of $p_{X_{n+1}|Y_{0:n+1}}(x_{n+1}; y_{0:n+1})$ w.r.t. x_{n+1} is equal to 1.

$$p_{x_{n+1}|Y_{0:n+1}}(x_{n+1}; y_{0:n+1}) \propto h(x_{n+1}; y_{0:n+1})$$

where $h(x_{n+1}; y_{0:n+1})$ is known. To compute $p_{x_{n+1}|y_{0:n+1}}(x_{n+1}; y_{0:n+1})$ we use that $\inf p_{X_{0:n+1}|Y_{0:n+1}} dx = 1$ and write:

$$p_{X_{n+1}|Y_{0:n+1}}(x_{n+1}; y_{0:n+1}) = \frac{h(x_{n+1}; y_{0:n+1})}{\int h(x; y_{0:n+1}) dx}$$

Generally speaking, the prediction and update expressions are intractable; however, there are two important cases in which closed-form expressions may be obtained. First, the linear Gaussian case leads to the Kalman filter, which is discussed in section 3.4. The second case arises when X_n takes its values in a finite state set, which will be discussed in section 3.5.

3.4. Gaussian linear case: Kalman algorithm

3.4.1. *Kalman filter*

Let us consider the model defined for $n \geq 0$ by the two following equations:

$$\begin{cases} X_{n+1} = A_n X_n + B_n \text{ (evolution equation)} \\ Y_n = C_n X_n + U_n \text{ (observation equation)} \end{cases} \quad [3.9]$$

where $\{A_n\}$ and $\{C_n\}$ are two sequences of matrices with adequate dimensions. In this context, A_n is called the state matrix and C_n the observation matrix.

$\{B_n\}$ and $\{U_n\}$ are two centered Gaussian vector sequences, independent of each other, with the covariances R_n^B and R_n^U, respectively. We also assume that the random vector X_0 is Gaussian with zero mean and covariance Σ_0.

The first equation of [3.9] describes the recurrence of the hidden state X_n. It is called the *state equation* or *evolution equation*.

The vector Y_n denotes the *measurement vector* and the second equation of [3.9] is called the *observation equation*.

Generally speaking, we wish to make inferences concerning states based on the observations. In this context, the evolution equation may be viewed as an *a priori* probability distribution of the states. For this reason, we can consider that we are in a Bayesian framework.

Let us return to equations [3.9]. As an exercise, the following properties can be shown:

– based on the linear transformation property of the Gaussian distribution, $(X_{0:n}, Y_{0:n})$ is jointly Gaussian;

– the sequence X_n is a Markov chain;

– the probability density of X_{n+1} given X_n is written as:

$$p_{X_{n+1}|X_n}(x_{n+1}, x_n) \sim \mathcal{N}(A_n X_n; R_n^B)$$

– the probability distribution of $Y_{0:n}$ conditionally on $X_{0:n}$ verifies the property of independence expressed in equation [3.2];

– the probability density of Y_n given X_n is written as:

$$p_{Y_n|X_n}(x_n, y_n) \sim \mathcal{N}(C_n X_n; R_n^U)$$

The bivariate process $(X_{0:n}, Y_{0:n})$ is therefore an HMM.

The fact that the probability distribution of X_n conditional on $Y_{0:n}$ is Gaussian constitutes a fundamental property. We therefore simply need to determine the expression of its mean and covariance. The Kalman algorithm provides a recursive way of calculating these two quantities. In this context, the following notation is commonly used:

$$X_{n|k} = \mathbb{E}\{X_n | Y_{0:k}\} \qquad [3.10]$$

$$P_{n|k} = \text{cov}(X_n | Y_{0:k}) \qquad [3.11]$$

Note that, in accordance with property 1.11, the Gaussian character implies that the conditional expectation $X_{n|k}$ corresponds to the orthogonal projection of X_n onto the linear space spanned by $Y_0, \ldots, Y_k$. The Kalman filter can therefore be deduced on the basis of geometric arguments alone, associated with the projection theorem (see section 1.3). In this context, note that $(X, Y) = \mathbb{E}\{XY\}$ denotes the scalar product of X and Y, and that $(X|Y_{0:k})$ is the orthogonal projection of X onto the linear space spanned by $Y_{0:k}$.

Before presenting the Kalman algorithm, let us consider a classic example, concerning trajectography, leading to equations of the form set out in [3.9].

EXAMPLE 3.1 (1D rectilinear motion).– Consider a vehicle moving along a straight line at the constant speed v. The position at the time $n+1$ is given by $d_{n+1} = d_n + vT$, where T denotes the sampling period. This motion equation can also be written as:

$$\begin{cases} d_{n+1} = d_n + v_n T \\ v_{n+1} = v_n \end{cases}$$

with the initial conditions d_0 and v_0. Note that the second equation of the system is reduced to $v_n = v_0$ if we assume that the vehicle has a constant speed. But if we have little faith in this hypothesis of a constant speed, the possible variability can be modeled by adding a random variable b_n to v_n. This leads to:

$$\begin{cases} d_{n+1} = d_n + v_n T \\ v_{n+1} = v_n + b_n \end{cases}$$

Hence, the evolution of the couple (d_n, v_n) can be rewritten in matrix form:

$$\begin{bmatrix} d_{n+1} \\ v_{n+1} \end{bmatrix} = \begin{bmatrix} 1 & T \\ 0 & 1 \end{bmatrix} \begin{bmatrix} d_n \\ v_n \end{bmatrix} + \begin{bmatrix} 0 \\ 1 \end{bmatrix} b_n$$

Let us now assume that only the position d_n is observed at the output of a noised device delivering the value $y_n = d_n + u_n$, where u_n is a random process used as a model for the *measurement noise*.

If we let $X_n = \begin{bmatrix} d_n & v_n \end{bmatrix}^T$, and write the expression of the observation Y_n in vector form, we get:

$$\begin{cases} X_{n+1} = AX_n + B_n \\ \quad Y_n = CX_n + U_n \end{cases}$$

where $C = \begin{bmatrix} 1 & 0 \end{bmatrix}$. This expression has the form similar to [3.9].

3.4.1.1. *Kalman filter algorithm*

The Kalman algorithm offers a recursive solution to the HMM filtering problem in the linear Gaussian case, as defined in expressions [3.9]. Its recursive characteristic should be understood in the sense that the number of operations and the amount of memory required by the algorithm do not increase as the number of observations increases. The Gaussian and linear characteristics of the model mean that the a posteriori law of X_n given $Y_{0:n}$ is Gaussian, and is therefore characterized completely by its mean $X_{n|n}$ and covariance $P_{n|n}$.

Exercise 3.1 gives a detailed proof of the Kalman algorithm in the scalar case. For the vectorial case, a fully similar proof leads to the Algorithm 7.

Data: $Y_{n\geq 0}, A_{n\geq 0}, C_{n\geq 0}, R^B_{n\geq 0}, R^U_{n\geq 0}, \Sigma_0$
Result: for $n \geq 0$, $X_{n|n}, P_{n|n}, P_{n|n-1}, \ell_n$
Initialization:
$X_{0|0} = 0, P_{0|0} = \Sigma_0$
$\Gamma_0 = C_0 \Sigma_0 C_0^T + R^U_0$
$\ell_0 = -\frac{1}{2}(n_y \log(2\pi) + Y_0^T \Gamma_0^{-1} Y_0 + \log \det\{\Gamma_0\})$
for $n \geq 1$ **do**

$\quad X_{n|n-1} = A_{n-1} X_{n-1|n-1}$ (prediction) $\qquad$ [3.12]

$\quad P_{n|n-1} = A_{n-1} P_{n-1|n-1} A_{n-1}^T + R^B_{n-1}$ $\qquad$ [3.13]

$\quad \Gamma_n = C_n P_{n|n-1} C_n^T + R^U_n$ $\qquad$ [3.14]

$\quad K_n = P_{n|n-1} C_n^T \Gamma_n^{-1}$ (Kalman gain) $\qquad$ [3.15]

$\quad i_n = Y_n - C_n X_{n|n-1}$ (innovation) $\qquad$ [3.16]

$\quad X_{n|n} = X_{n|n-1} + K_n i_n$ (update) $\qquad$ [3.17]

$\quad P_{n|n} = (I_{n_x} - K_n C_n) P_{n|n-1}$ $\qquad$ [3.18]

$\quad \ell_n = \ell_n - \frac{1}{2}(n_y \log(2\pi) + i_n^T \Gamma_n^{-1} i_n + \log \det\{\Gamma_n\})$ $\qquad$ [3.19]

end

/* n_x, n_y denote the dimensions of X_n and Y_n respectively. $X_{n|n}$ the filter outputs, $P_{n|n}$ the covariance of $X_{n|n}$, ℓ the log-likelihood of the observations. */

Algorithm 7: Kalman algorithm

The Kalman algorithm performs a computation of $X_{n|n}$ at the time n from the value $X_{n-1|n-1}$ in two successive steps: the *prediction* by equation [3.12] and the *update* by equation [3.17]. These two steps are similar to the two steps of the general cases in section 3.3. Moreover, this calculation only requires the memorization of the finite dimension state.

The sequence $i_n = Y_n - \mathbb{E}\{Y_n|Y_{0:n-1}\}$ is known as the *innovation*, and is the difference between that which is observed at instant n and that which can be "explained" by previous observations.

The sequence $\ell_n = 2\log p_{Y_{0:n}}(Y_{0:n})$ in algorithm 1 is the log-likelihood of the observations, calculated using the series of innovations. In the case where $p_{Y_{0:n}}(Y_{0:n})$ depends on an unknown parameter θ, the Kalman algorithm can be used to carry out maximization with respect to θ.

The expression of this algorithm calls for a few additive comments:

1) K_n is called the *Kalman gain*. It can be calculated beforehand, since it is determined by equations [3.13], [3.14] and [3.18], which do not depend on the observed data Y_n. However, except in the case of scalars (see exercise 3.1), the expression of K_n requires to solve a complex recursive equation referred to as the *Riccati equation*.

2) Let $R_n^U = 0$. Using expressions [3.14] and [3.15], we obtain $K_n = P_{n|n-1}C_n^T(C_n P_{n|n-1} C_n^T)^{-1}$. From this, we see that $C_n K_n = I_{n_y}$. In the absence of observation noise, the Kalman gain is the right inverse of the observation matrix.

3) Now, let $R_n^B = 0$. The Kalman gain can be shown to tend toward 0. In this case, the evolution model is highly reliable, and the estimation $X_{n|n}$ is principally calculated using the prediction $A_{n-1}X_{n-1|n-1}$.

4) The significance of equation [3.17] is obvious. According to the state equation of [3.9], the "best" value of X_n at time n is $A_{n-1}X_{n-1|n-1}$. Then, this value is corrected by a quantity proportional to the difference between what we observe, say Y_n, and what we can expect from the previous observation, say $C_n A_n X_{n-1|n-1}$. This may be seen as an adaptive compromise between what we see, i.e. Y_n, and what we know, i.e. the equation of evolution.

5) We wish to draw your attention to the fact that the algorithm requires us to know R_n^B and R_n^U. However, the theory can be generalized to include the situation where these quantities have to be estimated based on the observed data. In that case, the sequence of gains can no longer be calculated beforehand.

3.4.1.2. *Deterministic case*

Let us say a few words on the deterministic stationary case, which can be written as:

$$\begin{cases} X_{n+1} = AX_n \\ Y_n = CX_n \end{cases} \quad [3.20]$$

where $A_n = A$ and $C_n = C$.

An interesting issue is to say if we are able to perform X_0 from d_X observations $Y_{0:d_X-1}$, where d_X denotes the dimension of the state. In the literature, this property is known as *observability*. It is worth noting that if we can perform X_0, we can also perform X_n for any $n > 0$ using the first equation.

PROPERTY 3.1.– The system described by equation [3.20] is observable if and only if the following matrix $\mathcal{W}$ is of rank d_X:

$$\mathcal{W} = \begin{bmatrix} C \\ CA \\ \vdots \\ CA^{d_X-1} \end{bmatrix} \qquad [3.21]$$

PROOF.– Using [3.20], we can successively write:

$$Y_0 = CX_0$$
$$Y_1 = CAX_0$$
$$\vdots$$
$$Y_{d_X-1} = CA^{d_X-1}X_0$$

In more compact form, this set of linear equations with respect to X_0 can be written as $\mathcal{W}X_0 = \mathcal{Y}$, where $\mathcal{Y} = \begin{bmatrix} Y_0 & \dots & Y_{d_X-1} \end{bmatrix}^T$. Hence, there is a unique solution if and only if the matrix $\mathcal{W}$ is of rank d_X. Let us note that $\mathcal{W}$ is of dimension $(d_Y d_X, d_X)$, where d_Y is the dimension of Y. ∎

Non-observability does not mean that X_0 cannot be performed. This only means that the solution is not unique. It is reasonable to assume observability. If non-observability occurs, some components of the estimated hidden states could be large.

EXERCISE 3.1 (Kalman filter derivation, scalar case (see p. 208)).– Consider the system described by the two equations:

$$\begin{cases} X_{n+1} = a_n X_n + B_n \\ Y_n = c_n X_n + U_n \end{cases} \qquad [3.22]$$

where $\{a_n\}$ and $\{c_n\}$ are two sequences of scalars. $\{B_n\}$ and $\{U_n\}$ are two centered Gaussian sequences, independent of each other, with the variances $\sigma_B^2(n)$ and $\sigma_U^2(n)$, respectively. In particular, $\mathbb{E}\{B_n Y_k\} = 0$ and $\mathbb{E}\{U_n X_k\} = 0$.

We wish to determine the filtering distribution, i.e. the conditional distribution of X_n given $Y_{0:n}$. As we saw in section 1.4.3, this conditional distribution is Gaussian. Its mean corresponds to the orthogonal projection of X_n onto the sub-space spanned by $Y_{0:n}$ (expression [1.44]) and its covariance is given by [1.45]. Using the notation introduced in section 1.4.3, we can therefore write that $X_{n|n} = (X_n|Y_{0:n})$, $X_{n+1|n} = (X_{n+1}|Y_{0:n})$ and $\mathbb{E}\{Y_{n+1}|Y_{0:n}\} = (Y_{n+1}|Y_{0:n})$.

1) Show that $X_{n+1|n} = a_n X_{n|n}$;

2) Show that $(Y_{n+1}|Y_{0:n}) = c_{n+1} X_{n+1|n}$;

3) Use this result to deduce that K_n exists, such that $X_{n+1|n+1} = X_{n+1|n} + K_{n+1} i_{n+1}$, where $i_{n+1} = Y_{n+1} - c_{n+1} X_{n+1|n}$;

4) Let $P_{n+1|n} = (X_{n+1} - X_{n+1|n}, X_{n+1} - X_{n+1|n})$. Show that:

$$\mathrm{var}\,(i_{n+1}) = c_{n+1}^2 P_{n+1|n} + \sigma_U^2(n+1)$$

5) Show that the r.v. $i_{0:n}$ are independent. Use this to deduce the expression of $p_{Y_{0:n}}(y_{0:n})$ as a function of $i_{0:n}$ and $\mathrm{var}\,(i_{0:n})$;

6) From this result, deduce that:

$$K_{n+1} = \frac{P_{n+1|n} c_{n+1}}{\sigma_U^2(n+1) + c_{n+1} P_{n+1|n} c_{n+1}}$$

7) Show that:

$$P_{n+1|n} = a_n^2 P_{n|n} + \sigma_B^2(n)$$

then that:

$$P_{n+1|n+1} = (1 - K_{n+1} c_{n+1}) P_{n+1|n}$$

Bringing together all of the previous results, verify that we obtain algorithm 7.

EXERCISE 3.2 (Denoising an AR-1 using Kalman (see p. 228)).– We consider, $n \geq 0$, the discrete-time signal $Y_n = X_n + U_n$, where X_n denotes an zero-mean AR-1 process. This can be written as:

$$\begin{cases} X_{n+1} = aX_n + B_n & \text{(state equation)} \\ Y_n = X_n + U_n & \text{(observation equation)} \end{cases}$$

B_n and U_n are assumed to be two white uncorrelated noises. Let $\rho = \sigma_b^2/\sigma_u^2$. We want estimate X_n using the observed $Y_{0:n}$. Such a process is called AR-1 (auto-regressive of order 1).

Let us note that, provided that $|a| < 1$, the state equation has a unique solution which is a second-order stationary process whose expression is causal with respect to B_n [BLA 14]. It is easy to show that $\mathbb{E}\left\{X_n^2\right\} = \frac{\sigma_b^2}{1-a^2}$. This value can be used as the initial value $P_{0|0}$.

Let us note that the function scipy.signal.lfilter with the parameter (1,alpha) implements the equation $X_{n+1} + \alpha X_n = B_n$. Therefore, alpha=-a.

1) Determine, as a function of a and ρ, the recursive equation of the Kalman gain, as well as its initial value K_0. Show that the Kalman gain tends toward a limit, and determine the expression of this limit;

2) Write a program to implement the Kalman filter for this model.

EXERCISE 3.3 (Kalman filtering of a noisy 1D trajectory (see p. 230)).– Let us consider the 1D trajectory of length $N = 100$, generated by the first equation of the system model of example 3.1, which can be written as:

$$\begin{cases} X_{n+1} = AX_n + Db_n \\ Y_n = CX_n + U_n \end{cases} \quad [3.23]$$

where

$$A = \begin{bmatrix} 1 & 0.1 \\ 0 & 1 \end{bmatrix} \quad \text{and} \quad D = \begin{bmatrix} 0 \\ 1 \end{bmatrix}$$

b_n is a zero-mean Gaussian sequence with standard deviation equal to $\sigma_b = 1.0$, and U_n a zero-mean Gaussian sequence with standard deviation 10.0.

Write a program to (i) generate the state sequence and the observation sequence and (ii) estimate the state sequence from the observation using the Kalman filter. You will consider two cases: the first with $C = \begin{bmatrix} 1 & 0 \end{bmatrix}$ and the other with $C = \begin{bmatrix} 0 & 1 \end{bmatrix}$.

What do you observe? To conclude, perform the rank of $\mathcal{W}$ given by [3.21].

EXERCISE 3.4 (Calculating the likelihood of an ARMA (see p. 233)).– Consider the HMM model defined by:

$$\begin{cases} X_{n+1} = AX_n + RZ_{n+1} \\ Y_n = CX_n \end{cases} \quad [3.24]$$

where Z_n is a sequence of independent random variables, with a Gaussian distribution, of mean 0 and variance σ^2, and X_n is a vector of length r, where

$$A = \begin{bmatrix} -a_1 & 1 & 0 & \cdots & 0 \\ -a_2 & 0 & 1 & \cdots & 0 \\ \vdots & & & & \\ -a_{r-1} & 0 & 0 & \cdots & 1 \\ -a_r & 0 & 0 & \cdots & 0 \end{bmatrix}, \quad R = \begin{bmatrix} 1 \\ b_1 \\ b_2 \\ \vdots \\ b_{r-1} \end{bmatrix}, \quad C = \begin{bmatrix} 1 & 0 & 0 & \cdots & 0 \end{bmatrix}$$

1) Show that Y_n verifies the recursive equation:

$$Y_n + \sum_{m=1}^{r} a_m Y_{n-m} = Z_n + \sum_{k=1}^{r-1} b_k Z_{n-k}$$

Consequently, if $a_j = 0$ for $j > p$ and $b_j = 0$ for $j > q$, and taking $r = \max(p, q+1)$, Y_n verifies:

$$Y_n + \sum_{m=1}^{p} a_m Y_{n-m} = Z_n + \sum_{k=1}^{q} b_k Z_{n-k} \qquad [3.25]$$

Thus, if $a(z) = 1 + \sum_{m=1}^{p} a_m z^{-m} \neq 0$ for $|z| \geq 1$, Y_n is an ARMA-(p,q) [BLA 14], which is expressed causally as a function of Z_n.

2) Use the Kalman algorithm to calculate, recursively, the log-likelihood $\ell_n = \log p_{Y_{0:n}}(y_{0:n}; \theta)$ of an ARMA associated with parameter $\theta = \{a_{1:p}, b_{1:q}, \sigma^2\}$.

3) Let $\theta = \{a_{1:p}, b_{1:q}, \sigma^2\}$ and $\widetilde{\theta} = \{a_{1:p}, b_{1:q}, 1\}$. Determine the relationship giving the likelihood ℓ associated with θ as a function of the likelihood $\widetilde{\ell}$ associated with $\widetilde{\theta}$.

4) Write a function that uses $a_1, \ldots, a_p, b_1, \ldots, b_q$ and σ^2 to calculate the first n covariance coefficients of an ARMA-(p,q). First, let us note that if $\widetilde{Y}_n$ denotes the AR-p process defined by:

$$\widetilde{Y}_n + \sum_{m=1}^{p} a_m \widetilde{Y}_{n-m} = Z_n \qquad [3.26]$$

then the process Y_n defined by [3.25] is related to $\widetilde{Y}_n$ by

$$Y_n = \widetilde{Y}_n + \sum_{k=1}^{q} b_k \widetilde{Y}_{n-k} \qquad [3.27]$$

To calculate the covariance of $\widetilde{Y}_n$, use the following recursion [BRO 90]:

$$R(k) + a_1 R(k-1) + \ldots + a_p R(k-p) = 0 \quad \text{for } k \geq p+1 \qquad [3.28]$$

For the $(p+1)$ initial values, let us use:

$$\begin{bmatrix} R(0) & R(1) & \ldots & R(p-1) \\ R(1) & R(0) & & \\ \vdots & & & \\ R(p-1) & \ldots & \ldots & R(0) \end{bmatrix} \begin{bmatrix} a_1 \\ \vdots \\ a_p \end{bmatrix} = - \begin{bmatrix} R(1) \\ \vdots \\ R(p) \end{bmatrix} \qquad [3.29]$$

and $\sigma^2 = R(0) + a_1 R(1) + \ldots + R(p) a_p$. Equation [3.29] can be rewritten as a linear expression with respect to $R(0), R(1), \ldots, R(p-1)$. Once $R(k)$ is performed for k ranging from 0 to $n+r-1$, we can use the relation [3.27] to derive the following expression for the calculation of the covariance $\gamma(k)$ of the ARMA process:

$$\gamma_k = \sum_{i=0}^{q+1} \sum_{j=0}^{q+1} b_i b_j R(|k-i-j|) \qquad [3.30]$$

where $b_0 = 1$. Write a program to calculate the likelihood of an ARMA-(p,q) using the Kalman algorithm. Compare this result with those obtained using the direct calculation program.

3.4.2. *RTS smoother*

In general cases, smoothing consists of calculating $p_{X_n|Y_{0:N}}(x_n|y_{0:N})$, where N is the final observation time. Thus, this a "batch" (off-line) algorithm. It is shown that the HMM structure leads to recursive two-pass algorithms, one pass forward and the other backward [CAP 05]. Here we only give without proof the algorithm known as RTS (Rauch-Tung-Striebel) smoother.

Data: $Y_{n=0:N}$, $A_{n=0:N}$, $C_{n=0:N}$, $R^B_{n=0:N}$, $R^U_{n=0:N}$, Σ_1
Result: for $n = 0:N$, $X_{n|N}$, $P_{n|N}$
Initialization:
Call Kalman filter providing for $n = 0:N$, $X_{n|n}$, $P_{n|n}$, $P_{n|n-1}$ (see algorithm 7),
for $n = N-1$ to 1 *(backward pass)* **do**

$$K_n = P_{n|n} A_n^T P_{n+1|n}^{-1} \qquad [3.31]$$
$$X_{n|N} = X_{n|n} + K_n (X_{n+1|N} - A_n X_{n|n}) \qquad [3.32]$$
$$P_{n|N} = P_{n|n} - K_n (A_n P_{n|n} - P_{n+1|N} K_n^T) \qquad [3.33]$$

end

Algorithm 8: RTS smoother.

EXERCISE 3.5 (Filtering and smoothing for 2D tracking (see p. 237)).– Consider a mobile element in the plane Ox_1x_2. Let $x_1(t)$ and $x_2(t)$ be continuous time functions representing the two components of its position, $\dot{x}_1(t)$ and $\dot{x}_2(t)$ their first derivatives (speeds) and $\ddot{x}_1(t)$ and $\ddot{x}_2(t)$ their second derivatives (accelerations).

If the acceleration is null, i.e. $\ddot{x}_1(t) = 0$ and $\ddot{x}_2(t) = 0$, the trajectory of the mobile element is a straight line. When the acceleration is non-null, this may mean that the vehicle has increased or decreased its speed in a straight line, and/or that the vehicle has deviated from the straight line.

To take account of possible acceleration, $\ddot{x}_1(t)$ and $\ddot{x}_2(t)$ are modeled by two Gaussian, centered white noises of the same variance σ^2. This allows us to track a mobile element with a trajectory that does not follow a straight line.

Let T be the sampling period. For $i = 1$ and $i = 2$, let $X_{i,n} = x_i(nT)$ and $\dot{X}_{i,n} = \dot{x}_i(nT)$, and let $X_n = \begin{bmatrix} X_{1,n} & X_{2,n} & \dot{X}_{1,n} & \dot{X}_{2,n} \end{bmatrix}^T$.

We presume that only the position is observed, and that this observation is subject to additional noise. The position is therefore written as $Y_n = CX_n + U_n$ with:

$$C = \begin{bmatrix} 1 & 0 & 0 & 0 \\ 0 & 1 & 0 & 0 \end{bmatrix}$$

and U_n is the observation noise.

1) Use a first-order Taylor approximation to show that:

$$X_{n+1} = AX_n + B_n \text{ with } A = \begin{bmatrix} 1 & 0 & T & 0 \\ 0 & 1 & 0 & T \\ 0 & 0 & 1 & 0 \\ 0 & 0 & 0 & 1 \end{bmatrix} \quad [3.34]$$

and where B_n is a noise. Give the covariance matrix as a function of T and σ^2.

2) Suppose that $T = 1/10$ s and the speed is of the order of 30 m/s (approximately 90 km/h). Explain the connection between σ and a possible acceleration over a duration T. The speed may be considered to vary by a quantity proportional to v_0.

3) Write functions implementing the Kalman filter algorithm and the RTS smoothing algorithm in the case where the parameters of the model are not dependent on time. Write a program to simulate a 2D trajectory, compute filtering and smoothing, and compare their confidence ellipses. Note that the confidence region at $100\alpha\%$ of a random Gaussian vector of dimension 2, mean μ and covariance matrix C is an ellipse with equation $(x-\mu)^T C^{-1}(x-\mu) = -2\log(1-\alpha)$.

To create a 2D trajectory, you are not obliged to use equation [3.34], you can take of any form, for example that obtained using:

```
# -*- coding: utf-8 -*-
""" 
Created on Wed Jun 29 10:51:12 2016
****** generate2Dtrajectory
@author: maurice
"""
from numpy import cos, sin, arange, pi
from matplotlib import pyplot as plt
MO = 2; N = 50; t = 2*pi*arange(0,N)/N/4.0;
X1 = cos(t); X2 = cos(t)**2+sin(t)
plt.clf(); plt.plot(X1,X2); plt.show()
```

We see that the plotted trajectory consists of two almost linear parts and a highly curved part. It is clear that the linear parts can be better modeled by equation [3.34] than the curved part. It follows that better filtering and smoothing can be expected in the linear parts.

3.5. Discrete finite Markov case

In cases where the states X_n of an HMM take their value in a finite set of values, the inference on X_n has a closed-form expression. This expression is based on two smoothing algorithms, as defined in section 3.2, known as the Baum-Welch or forward-backward algorithms.

Consider an HMM with hidden states X_n, where n ranges from 0 to $N-1$, which have values in a finite set $S = \{0, 2, \ldots, S-1\}$. The distribution of the observations Y_n given $X_n = i$ is taken to have a probability density denoted as $g_n(y_n|i)$. Let us denote $p_n(i|j) = \mathbb{P}\{X_n = i|X_{n-1} = j\}$ and $\omega(i) = \mathbb{P}\{X_0 = i\}$.

EXERCISE 3.6 (Discrete HMM generation (see p. 242)).– Consider an HMM with $S = 4$ hidden states, with an initial distribution $\omega_0 = \begin{bmatrix} 1/2 & 1/4 & 1/8 & 1/8 \end{bmatrix}$ and the following transition probability matrix[1]:

$$P = \begin{bmatrix} 0.4 & 0.1 & 0.3 & 0.2 \\ 0.1 & 0.4 & 0.3 & 0.2 \\ 0.3 & 0.1 & 0.4 & 0.2 \\ 0.1 & 0.3 & 0.1 & 0.5 \end{bmatrix} \quad \text{where} \quad P_{j,i} = p(i|j)$$

---

[1] The number in line j denotes the initial state, and the number in column i gives the final state. Thus, $\sum_i P_{j,i} = 1$, i.e. the sum of the elements in a line is equal to 1.

Let $F_s(x)$ be the cumulative function associated with the conditional distribution of X_n given $X_{n-1} = s$. More precisely:

$$F_s(x) = \sum_{i=0}^{S-1} \mathbb{1}(x \leq i) p(i|s) = \sum_{i=0}^{S-1} \mathbb{1}(x \leq i) P_{s,i}$$

1) Show that:

$$X_{n+1} = \sum_{j=0}^{S-1} j \times \mathbb{1}(U_n \in [F_{X_n}(j-1), F_{X_n}(j)])$$

where U_n is a sequence of independent r.v.s with a uniform distribution over the interval $(0, 1)$.

2) Consider that the observations Y_n, conditionally on the states X_n, are Gaussian random vectors of size d, with respective means μ_i and covariance C_i with $i = 0$ to $S - 1$. Using the results of exercise 4.2, write a function to generate a sequence of data following the proposed HMM.

For $d = 2$, plot the vectors Y_n and their ellipse of confidence at 95%.

We suggest for generating the positive covariance matrix C_i to draw an auxiliary random matrix M_i and execute $C_i = M_i M_i^T$.

3.5.1. *Forward-backward formulas*

Now let us determine two recursive formulas, known as the *forward-backward* formulas. We will then consider the way in which these formulas are used in 1 and 2 instant smoothing algorithms, and in the algorithm used to estimate the hidden variables $X_{0:N}$ on the basis of the observations $Y_{0:N}$.

Let:

$$\alpha_n(i) = \mathbb{P}\{X_n = i | Y_{0:n}\} \qquad [3.35]$$

$$\beta_n(i) = p_{Y_{n+1:N}|X_n=i}(y_{n+1:N}|X_n = i) \frac{L(y_{0:n})}{L(y_{0:N})} \qquad [3.36]$$

where

$$L(y_{0:n}) = p_{Y_{0:n}}(y_{0:n}) \qquad [3.37]$$

is the likelihood associated with the observations $Y_{0:n}$ assuming that the distribution of $Y_{0:n}$ has a density. In the case of discrete r.v.s, $p_{Y_{0:n}}(y_{0:n})$ should be replaced by $\mathbb{P}\{Y_{0:n} = y_{0:n}\}$.

3.5.1.1. *Forward recursion*

The recursion formula giving $\alpha_n(i)$ as a function of $\alpha_{n-1}(i)$ can be written as:

$$\alpha_n(i) = \frac{\widetilde{\alpha}_n(i)}{c_n} \quad [3.38]$$

with

$$\widetilde{\alpha}_n(i) = g_n(y_n|i) \times \sum_{j=0}^{S-1} \alpha_{n-1}(j) p_n(i|j) \text{ and } c_n = \sum_{i=0}^{S-1} \widetilde{\alpha}_n(i) \quad [3.39]$$

We also derive the log-likelihood recursion:

$$\ell_N = \log L(y_{0:N}) = \sum_{n=0}^{N} \log c_n \quad [3.40]$$

PROOF.– We can write:

$$\mathbb{P}\{X_n = i|Y_{0:n}\} = \sum_{j=0}^{S-1} \mathbb{P}\{X_n = i, X_{n-1} = j|Y_{0:n}\} \quad [3.41]$$

Based on expression [3.3], we have:

$$\mathbb{P}\{X_{0:n} = x_{0:n}|Y_{0:n}\} L(y_{0:n}) = \mathbb{P}\{X_{0:n-1} = x_{0:n-1}|Y_{0:n-1}\} L(y_{0:n-1}) \times \ldots$$
$$p_n(x_n|x_{n-1}) \times g_n(y_n|x_n)$$

Summing on x_k with $k = 0$ to $n - 2$, we obtain:

$$\mathbb{P}\{X_{n-1} = x_{n-1}, X_n = x_n|Y_{0:n}\} = \frac{L(y_{0:n-1})}{L(y_{0:n})} \times \ldots$$
$$\mathbb{P}\{X_{n-1} = x_{n-1}|Y_{0:n-1}\} \times p_n(x_n|x_{n-1}) \times g_n(y_n|x_n)$$

Applying this expression to [3.41], we obtain:

$$\alpha_n(i) = g_n(y_n|i) \times \frac{L(y_{0:n-1})}{L(y_{0:n})} \times \sum_{j=0}^{S_1} \alpha_{n-1}(j) p_n(i|j) \quad [3.42]$$

Let us note that the initial value is written as $\alpha_0(i) = g_0(Y_0|i)\,\omega_0(i)/L(y_0)$.

REMARK: the sum of the $\alpha_n(i)$ for i ranging from 0 to $S-1$ is equal to 1. For this reason, we simply need to calculate the following terms:

$$\widetilde{\alpha}_n(i) = g_n(y_n|i) \times \sum_{j=0}^{S-1} \alpha_{n-1}(j) p_n(i|j)$$

and then sum these terms over i in order to obtain the normalization constant. This constant is expressed as:

$$c_n = \frac{L(y_{0:n})}{L(y_{0:n-1})} = \sum_{i=0}^{S-1} \widetilde{\alpha}_n(i) \quad [3.43]$$

This result allows us to find an important formula giving the log-likelihood of the N observations: $\ell_N = \log L(y_{0:N}) = \sum_{n=0}^{N} \log c_n$. ∎

In summary, the forward algorithm is written as:

Data: $n = 0$ to N, $y_n \in \mathbb{R}^d$,
$\quad i, j = 0$ to $S - 1$, $\omega_0(i)$, $g_n(y|i)$, $p_n(i|j)$
Result: $\alpha_n(i)$, c_n, ℓ_n
Initialization:
for $i = 0$ *to* $S - 1$ **do**
$\quad | \quad \widetilde{\alpha}_0(i) = g_1(y_0|i)\omega_0(i);$
end
$c_1 = \sum_{i=0}^{S-1} \widetilde{\alpha}_1(i);\ \ell_0 = \log c_0;$
for $i = 0$ *to* $S - 1$ **do**
$\quad | \quad \alpha_0(i) = \widetilde{\alpha}_0(i)/c_0;$
end
for $n = 1$ *to* N **do**
$\quad$ **for** $i = 0$ *to* $S - 1$ **do**
$\quad\quad | \quad \widetilde{\alpha}_n(i) = g_n(y_n|i) \sum_{j=0}^{S-1} \alpha_{n-1}(j) p_n(i|j);$
$\quad$ **end**
$\quad c_n = \sum_{i=0}^{S-1} \widetilde{\alpha}_n(i);$
$\quad$ **for** $i = 0$ *to* $S - 1$ **do**
$\quad\quad | \quad \alpha_n(i) = \widetilde{\alpha}_n(i)/c_n;$
$\quad$ **end**
$\quad \ell_n = \ell_{n-1} + \log c_n;$
end

Algorithm 9: Forward recursion

3.5.1.2. Backward recursion

We now wish to determine the recursion giving $\beta_n(i)$ as a function $\beta_{n+1}(i)$. As an exercise, following an approach very similar to that used to obtain [3.38], we can show that:

$$\beta_n(i) = \frac{L(y_{0:n})}{L(y_{0:n+1})} \times \sum_{j=0}^{S-1} \beta_{n+1}(j) p_{n+1}(j|i) g_{n+1}(y_{n+1}|j) \qquad [3.44]$$

with the final value $\beta_N(i) = 1$ for any i. The backward algorithm is written as:

Data: $n = 0$ to N, $i, j = 0$ to $S - 1$,
$Y_n \in \mathbb{R}^d$, $g_n(y|i)$, $p_n(i|j)$,
c_n (from algorithm 9)
Result: $\beta_n(i)$
Initialization:
for $i = 0$ to $S - 1$ **do**
| $\beta_N(i) = 1$;
end
for $n = N - 1$ to 0 **do**
| **for** $i = 0$ to $S - 1$ **do**
| | $\beta_n(i) = \dfrac{1}{c_{n+1}} \sum_{j=0}^{S-1} \beta_{n+1}(j) p_{n+1}(j|i) g_{n+1}(y_{n+1}|j)$;
| **end**
end

Algorithm 10: Backward recursion

3.5.2. Smoothing formula at one instant

In this section, for the sake of simplicity, expression $\mathbb{P}\{X_{0:N} = x_{0:N} | Y_{0:N}\} \times L(y_{0:n})$ will be noted $\mathbb{P}\{X_{0:N} = x_{0:N}, Y_{0:N} = y_{0:N}\}$, which is more concise, but only correct in cases where X_n and Y_n are random variables with discrete values. Note that $L(y_{0:n})$ is obtained using expression [3.37]. Let us denote:

$$\gamma_n(i) = \mathbb{P}\{X_n = i | Y_{0:N}\} \qquad [3.45]$$

We have the following recursion:

$$\gamma_n(i) = \alpha_n(i) \times \beta_n(i) \qquad [3.46]$$

PROOF.– Using expression [3.4], we can write that:

$$\mathbb{P}\{X_{0:N} = x_{0:N}, Y_{0:N} = y_{0:N}\} = \mathbb{P}\{X_{0:n} = x_{0:n}, Y_{0:n} = y_{0:n}\} \ldots \quad [3.47]$$
$$\times \mathbb{P}\{X_{n+1:N} = x_{n+1:N}, Y_{n+1:N} = y_{n+1:N}\}/\mathbb{P}\{X_n = x_n\}$$

Summing over $x_{0:n-1}$ and $x_{n+1:N}$ and replacing x_n by i, we obtain:

$$\mathbb{P}\{X_n = i, Y_{0:N} = y_{0:N}\} = \mathbb{P}\{X_n = i, Y_{0:n} = y_{0:n}\} \ldots$$
$$\times \mathbb{P}\{Y_{n+1:N} = y_{n+1:N} | X_n = i\}$$

Using expressions [3.35], we obtain:

$$\mathbb{P}\{X_n = i, Y_{0:N} = y_{0:N}\} = \alpha_n(i) \times \beta_n(i) \times L(y_{0:N})$$

Therefore, [3.46]. ∎

3.5.3. *Smoothing formula at two successive instants*

Let us denote:

$$\xi_n(i,j) = \mathbb{P}\{X_{n+1} = i, X_n = j | Y_{0:N}\} \quad [3.48]$$

We have:

$$\xi_n(i,j) = \alpha_n(j)\beta_{n+1}(i)p_{n+1}(i|j)g_{n+1}(y_{n+1}|i) \quad [3.49]$$

PROOF.– Starting with expression [3.6], it may be rewritten as:

$$p_{X_{n:n+1},Y_{0:N}}(x_n, x_{n+1}, y_{0:N}) = p_{X_n|Y_{0:n}}(x_n, y_{0:n})$$
$$\underbrace{p_{Y_{n+1}|X_{n+1}}(x_{n+1}, y_{n+1})}_{g_{n+1}(y_{n+1}|x_{n+1})} \underbrace{p_{X_{n+1}|X_n}(x_{n+1}, x_n)}_{p_{n+1}(x_{n+1}, x_n)} \underbrace{p_{Y_{n+2:N}|X_{n+1}}(x_{n+1}, y_{n+2:N})}_{\frac{L(y_0:N)}{L(y_0:n+1)}\beta_{n+1}(x_{n+1})}$$

Dividing by $L(y_{0:n})$ and using the definition of $\alpha_n(i)$, expression [3.43] and equation [3.43], which is written as $c_{n+1} = \frac{L(y_{0:n+1})}{L(y_{0:n})}$, lead to [3.49]. ∎

Bringing together the forward and backward algorithms along with expressions [3.46] and [3.49], we obtain a means of calculating smoothing formulas for cases with one and two instants.

3.5.4. *HMM learning using the EM algorithm*

As we shall see, the one and two instant smoothing formulas are required in order to calculate the auxiliary function of the EM algorithm associated with the estimation of $\theta = (\omega_i, p(i|j), \rho_i)$, where ρ_i is a parameter of the observation distribution $g(y|i) = g(y; \rho_i)$. In this case, we presume that these distributions are not dependent on n.

Let us prove that the auxiliary function of the EM algorithm, as defined in algorithm 5, associated with the distribution of the discrete HMM has the following expression:

$$Q(\theta, \theta') = \sum_{n=0}^{N} \sum_{i=0}^{S-1} \log g(y_n; \rho_i) \gamma'_n(i) + \sum_{n=0}^{N-1} \sum_{i=0}^{S-1} \sum_{j=0}^{S-1} \log p(i|j) \xi'_n(i,j)$$

$$+ \sum_{i=0}^{S-1} \log(\omega_i) \gamma'_1(i) \qquad [3.50]$$

where the prime indicates that the calculated quantity is associated with the value θ' of the parameter.

PROOF.– Indeed, the joint probability law of $(X_{0:N}, Y_{0:N})$ is written as:

$$p_{X_{0:N}, Y_{0:N}}(x_{0:N}, y_{0:N}) = \prod_{n=0}^{N} g(y_n | X_n = x_n) \times$$

$$\prod_{n=1}^{N} \mathbb{P}\{X_{n+1} = x_{n+1} | X_n = x_n\} \mathbb{P}\{X_0 = x_0\}$$

Taking its logarithm, we have:

$$\log p_{X_{0:N}, Y_{0:N}}(x_{0:N}, y_{0:N}) = \sum_{n=0}^{N} \log g(y_n | X_n = x_n)$$

$$+ \sum_{n=1}^{N} \log \mathbb{P}\{X_{n+1} = x_{n+1} | X_n = x_n\} + \log \mathbb{P}\{X_0 = x_0\}$$

Using the identity $f(x_k) = \sum_{i=0}^{S-1} f(i) \mathbb{1}(x_k = i)$, we obtain:

$$\log p_{X_{0:N}, Y_{0:N}}(x_{0:N}, y_{0:N}) = \sum_{j=0}^{S-1} \sum_{n=0}^{N} \log g(y_n|j) \mathbb{1}(X_n = i)$$

$$+ \sum_{i=0}^{S-1} \sum_{j=0}^{S-1} \sum_{n=0}^{N-1} \log p(i|j) \mathbb{1}(X_{n+1} = i, X_n = j) + \sum_{j=1}^{S} \log \omega_j \mathbb{1}(X_1 = j)$$

Taking the conditional expectation with respect to $Y_{0:N}$ under the parameter θ' and using the fact that $\mathbb{E}\left\{\mathbb{1}(X=i)\right\} = \mathbb{P}\left\{X=i\right\}$, we have:

$$\mathbb{E}_{\theta'}\left\{\mathbb{1}(X_n=i)|Y_{0:N}\right\} = \mathbb{P}_{\theta'}\left\{X_n=i|Y_{0:N}\right\} = \gamma'_n(i)$$

$$\mathbb{E}_{\theta'}\left\{\mathbb{1}(X_{n+1}=i, X_n=j)|Y_{0:N}\right\} = \mathbb{P}_{\theta'}\left\{X_{n+1}=i, X_n=j|Y_{0:N}\right\} = \xi'_n(i,j)$$

$$\mathbb{E}_{\theta'}\left\{\mathbb{1}(X_1=i)|Y_{0:N}\right\} = \mathbb{P}_{\theta'}\left\{X_0=i|Y_{0:N}\right\} = \gamma'_0(i)$$

This demonstrates [3.50]. ∎

3.5.4.1. *Re-estimation formulas*

The maximization of $Q(\theta, \theta')$ with respect to θ is carried out as follows. Canceling the first derivative with respect to ω_i, under the constraint that $\sum_i \omega_i = 1$, we obtain:

$$\omega_i = \frac{\gamma'_1(i)}{\sum_{j=0}^{S-1} \gamma'_1(j)} \qquad [3.51]$$

Canceling the first derivative of Q with respect to $p(i|j)$, under the constraint that $\sum_i p(i|j) = 1$ for any j, we have:

$$p(i|j) = \frac{\sum_{n=0}^{N-1} \xi'_n(i,j)}{\sum_{i=0}^{S-1} \sum_{n=0}^{N-1} \xi'_n(i,j)} \qquad [3.52]$$

In addition, we assume that the distribution $g(y|i)$ is Gaussian, with mean μ_i and covariance C_i. Canceling the first derivative of Q with respect to μ_i, we have:

$$\mu_i = \frac{\sum_{n=0}^{N} Y_n \gamma'_n(i)}{\sum_{n=0}^{N} \gamma'_n(i)} \qquad [3.53]$$

Similarly, canceling the first derivative of Q with respect to C_i, we have:

$$C_i = \frac{\sum_{n=0}^{N} \gamma'_n(i)(Y_n - \mu_i)(Y_n - \mu_i)^T}{\sum_{n=0}^{N} \gamma'_n(i)} \qquad [3.54]$$

EXERCISE 3.7 (EM algorithm for a HMM (see p. 245)).– Consider an HMM with $S=4$ discrete states, including the initial state distribution $\omega = \begin{bmatrix} 1/2 & 1/4 & 1/8 & 1/8 \end{bmatrix}$, with the following matrix of transition probabilities:

$$P = \begin{bmatrix} 0.4 & 0.1 & 0.3 & 0.2 \\ 0.1 & 0.4 & 0.3 & 0.2 \\ 0.3 & 0.1 & 0.4 & 0.2 \\ 0.1 & 0.3 & 0.1 & 0.5 \end{bmatrix}, \quad \text{where} \quad P_{j,i} = p(i|j)$$

and the densities $g(y; \mu_i, C_i)$ are Gaussian, with respective means μ_i and covariances C_i. Let us remark that μ_i and C_i are assumed to be independent of n. Let $\theta = (\mu_i, C_i, \omega_i, p(i|j))$ and $Y_{0:N}$ be a sequence of N observations.

1) Use *Python*® to write a function to implement algorithms [9] and [10]. The inputs consist of the observations, along with ω, $p(i|j)$, μ_i and C_i. The function will perform the sequences α and β and the likelihood.

2) Write a function to estimate θ using the EM algorithm.

3) Test this algorithm using the generator obtained in exercise 3.6.

3.5.5. *The Viterbi algorithm*

Consider an HMM of which the states have values in a finite set $S = \{0, \ldots, S-1\}$ of S values. The transition distributions $p_n(i|j) = \mathbb{P}\{X_n = i | X_{n-1} = j\}$ are assumed to be known, as are the probability densities Y_n conditionally on $X_n = i$, denoted by $g_n(y|i)$.

We observe $y_0, \ldots, y_{N-1}$ and we wish to determine the sequence $x_0, \ldots, x_{N-1}$ that maximizes $\mathbb{P}\{X_{0:N-1} = x_{0:N-1} | y_{0:N-1}\}$. Maximizing $\{X_{0:N-1} = x_{0:N-1} | y_{0:N-1}\}$ with respect to $x_{0:N}$ is equivalent to maximizing the joint distribution of $(X_{0:N-1}, Y_{0:N-1})$.

The "brute force" approach involves calculating the joint law for the S^N possible configurations of the sequences $x_{0:N-1}$. As we shall see, the Viterbi algorithm reduces the number of calculations requiring only NS^2 steps, which is considerably lower than S^N.

We assume that, at time step $n-1$, we have S optimal sequences of length $n-1$ ending with the S possible values of x_{n-1}. In what follows, the sequence ending with the value $j \in \{0, \ldots, S-1\}$ at instant $(n-1)$ will be referred to as the jth path of length n. Let $\mathrm{met}_{n-1}(j)$ be the associated joint probability, known as the path metric. Using [3.3], taking the logarithm and noting $d_n(i|j) = \log \mathbb{P}\{X_{0:n-2} = x_{0:n-2}, X_{n-1} = j, X_n = i, y_{0:n}\}$, we have:

$$d_n(i|j) = \mathrm{met}_{n-1}(j) + \log g_n(y_n|i) + \log p_n(i|j)$$

where $d_n(i|j)$ is known as the branch metric (the branch ranging from j to i at step n). There are S possible ways of extending the jth path of length n. However, as we wish to find the maximum metric, only the ascendant giving the maximum metric should be saved. Consequently, the ith path of length $(n+1)$ has the following metric: $\mathrm{met}_n(i) = \max_{j \in \{0,\ldots,S-1\}} d_n(i|j)$. The ascendant giving the maximum value is written as $\mathrm{asc}_n(i) = \arg\max_{j \in \{0,\ldots,S-1\}} d_n(i|j)$ It must be saved in order to calculate the optimal sequence at final step N.

Figure 3.2 shows a calculation diagram for $S = 3$ and $N = 6$, with a graph of $SN = 18$ nodes which is known as a lattice. At step n, we calculate the $S^2 = 9$ branch metrics. From these nine possible extensions, we only retain the $S = 3$ optimal metrics reaching the S nodes of step n, along with their ascendants.

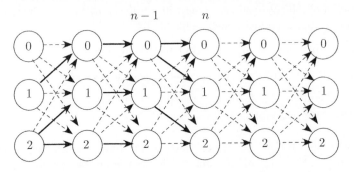

Figure 3.2. *Lattice with 3 states for a sequence of length 6. At step n, we keep the best ascendants (plain arrows) and compute the remaining three path metrics*

In conclusion, for each stage, we determine the S possible paths of length n with their associated metrics. The calculation is continued up to step N. The optimal sequence is obtained at the end of the process via backtracking.

Data: $g(Y(n), i), p(n, i, j)$,
for n from 0 to $N - 1$, for i and j from 0 to $S - 1$
Result: $X(n)$ for n from 0 to $N - 1$
Initialization: $\mathrm{met}(0, j) = 0$ for j from 0 to $S - 1$:
for $n = 0$ *to* $N - 1$ **do**
　for $i = 0$ *to* $S - 1$ **do**
　　for $j = 0$ *to* $S - 1$ **do**
　　　$d(i, j) = \mathrm{met}(n - 1, j) + \log g(n, Y(n), i) + \log p(n, i, j)$;
　　end
　end
　for $i = 0$ *to* $S - 1$ **do**
　　$\mathrm{met}(n, i) = \max_j d(i, j)$;
　　$\mathrm{asc}(n, i) = \arg\max_j d(i, j)$;
　end
end
$X(N - 1) = \arg\max_i \mathrm{met}(N - 1, i)$;
for $n = N - 2$ *to* 0 **by step** -1 **do**
　$X(n) = \mathrm{asc}(n + 1, X(n + 1))$;
end

Algorithm 11: Viterbi algorithm

The Viterbi algorithm is a dynamic programming algorithm in a lattice. In theory, the optimal sequence is deduced based on the totality of the N observations. In practice, if the observations are time indexed, stopping criteria are used in order to take intermediate decisions before the final observation.

EXERCISE 3.8 (State estimation by the Viterbi algorithm (see p. 249)).– Consider a discrete HMM with S states. Under the state s, the observation is Gaussian distributed with mean μ_s and variance σ_s^2. We denote P as the transition matrix and logG the matrix of the log-likelihoods associated with each hidden state. Write a program, in *Python*®, to generate a sequence of N observations, compute their associated log-likelihood matrix of size $N \times S$ and extract the sequence of states using algorithm 11.

4

Monte-Carlo Methods

4.1. Fundamental theorems

As we saw in section 1.6, theorems 1.8 and 1.9 form the basis for statistical methods and are crucial to the validity of Monte-Carlo methods. These theorems set out the way in which empirical means converge toward statistical moments. Noting that a statistical moment is defined as the integral of a certain function, this statement says, in some ways, that you can approximate this integral using a mean based on random (or pseudo-random) sequences. Using these two theorems, we see that the convergence as a function of the number N of samples is of the order of $N^{-1/2}$. It is therefore interesting to compare this value to those obtained using deterministic numerical methods, such as the trapezoid method or Simpson's method. The deterministic method can be seen to have a convergence speed of the order of $N^{-2/d}$, where d is the dimension of the space over which the function to integrate is defined. Consequently, Monte-Carlo methods present two advantages compared to deterministic methods, which are as follows: (i) the convergence speed does not depend on the dimension d, and (ii) their use does not depend on the regularity of the function being integrated.

In a less formal manner, the trapezoidal method can be seen as using a grid with a large number of points; many of them have a negligible effect on the calculated value of the integral; following the Monte-Carlo method, only the significant values are used. There is, however, one major drawback, in that the numerical result depends on the realization: the error is therefore random.

4.2. Stating the problem

The aim of Monte-Carlo methods is to calculate an integral using a random generator rather than a deterministic value set. The term Monte-Carlo refers to the role of chance in games played in the world-famous casino in Monaco. Let us consider an integral over $\mathbb{R}$, which is written as follows:

$$I(g) = \int_{-\infty}^{+\infty} g(x)dx \qquad [4.1]$$

If the integral is defined over $\mathcal{S} \subset \mathbb{R}$, it may be written using the indicative function of $\mathcal{S}$ in the following form:

$$\int_{\mathcal{S}} f(x)dx = \int_{-\infty}^{+\infty} \underbrace{f(x)\mathbb{1}_{\mathcal{S}}(x)}_{g(x)} dx$$

where $g(x) = f(x)\mathbb{1}_{\mathcal{S}}(x)$ and where $\mathbb{1}_{\mathcal{S}}(x)$ has a value of 1 if $x \in \mathcal{S}$, and 0 in all other cases. It therefore takes the form of expression [4.1].

In many applications, the integral for calculation is associated with the mathematical expectation of a function f, i.e. an integral of the following form:

$$I(g) = \int_{-\infty}^{+\infty} \underbrace{f(x)\,p_X(x)}_{g(x)} dx \qquad [4.2]$$

where $p_X(x)$ is the density of a probability distribution. Sometimes, the function to integrate does not have an explicit form and can only be calculated by using an algorithm.

The central idea behind the Monte-Carlo method is to use a random generator with a distribution characterized by a probability density μ, and then to use the law of large numbers to calculate the integral. We may write:

$$I(g) = \int_{-\infty}^{+\infty} g(x)dx = \int_{-\infty}^{+\infty} \frac{g(x)}{\mu(x)}\mu(x)dx = \int_{-\infty}^{+\infty} h(x)\mu(x)dx \qquad [4.3]$$

where $h(x) = g(x)/\mu(x)$. If we have a realization of N random variables $X_0, \ldots, X_{N_1}$ that are independent and identically distributed following distribution μ, we may arrive at the following expression:

$$I(g) \approx \frac{1}{N}\sum_{n=0}^{N-1} h(X_n) = \frac{1}{N}\sum_{n=0}^{N-1} \frac{g(X_n)}{\mu(X_n)}$$

More precisely, the law of large numbers states that:

$$\frac{1}{N} \sum_{n=0}^{N-1} \frac{g(X_n)}{\mu(X_n)} \xrightarrow[N \to +\infty]{} I(g)$$

where the convergence is in probability. Note that when the integral $I(g)$ is associated with a mathematical expectation as in expression [4.2], it is not, *a priori*, necessary to draw the sample using the distribution $p_X(x)$. As we shall see, the choice of a different distribution can even lead to a better approximation in these cases.

Note that a real random variable is a measurable function from a sample space Ω into $\mathbb{R}$, written $X : \omega \in \Omega \mapsto x \in \mathbb{R}$. Each outcome (or experiment) is associated with a real value x known as the realization. In the context of a statistical simulation, stating that N independent draws will be used signifies that N distinct random variables $X_0, \ldots, X_{N-1}$ will be considered, and that for each outcome $\omega \in \Omega$, we obtain N realizations $x_0, \ldots, x_{N_1}$, and not N realizations of a single random variable! In more general terms, we consider an infinite series of random variables, i.e. a family $\{X_n\}$ of r.v. indexed by $\mathbb{N}$. The practical applications of statistical methods are thus essentially linked to the properties obtained when N tends toward infinity. These properties are often easier to establish when the random variables in the series are considered as independent.

This approach is easily extended to the integral of a function defined over $\mathbb{R}^d$ and with values in $\mathbb{R}$. Let $g : (x^1, \ldots, x^d) \in \mathbb{R}^d \mapsto \mathbb{R}$ be a function of this type, and consider the following integral:

$$I(g) = \int_{-\infty}^{+\infty} \cdots \int_{-\infty}^{+\infty} g(x^1, \ldots, x^d) dx^1 \ldots dx^d$$

To calculate an approximate value for $I(g)$, we carry out N independent random drawings, for which the probability distribution defined over $\mathbb{R}^d$ has a probability density $\mu(x^1, \ldots, x^d)$, and we write:

$$I(g) \approx \frac{1}{N} \sum_{n=1}^{N} h(X_n^1, \ldots, X_n^d)$$

where $h(x^1, \ldots, x^d) = g(x^1, \ldots, x^d)/\mu(x^1 \ldots x^d)$.

Two types of problems should be considered when using the Monte-Carlo method to calculate an integral:

1) how to determine the "optimum" way of choosing the drawing distribution μ in order to calculate a given interval;

2) how to create samples following a given distribution.

We shall begin by considering the second problem, the generation of random variables. In this context, we shall begin by presenting distribution transformation methods, followed by sequential methods based on Markov chains. We shall then consider the first issue in a section on variance reduction.

First, knowledge of some history is required. The first method for calculating integrals by using a Monte-Carlo type technique was proposed by N. Metropolis in 1947, in the context of a statistical physics problem. In 1970, K. Hastings published an article establishing the underlying principle for general random variable generation methods, known as the Metropolis-Hastings sampler and based on Markov chains; in this context, we speak of Monte-Carlo Markov Chains (MCMC). In 1984, S. Geman and D. Geman proposed the "Gibbs" sampler, a specific form of the Metropolis-Hastings sampler, which was used by the authors in the context of image restoration.

4.3. Generating random variables

In this section, we shall consider that we have access to a generator using uniform distribution over the interval $(0,1)$, which is able to supply a given number of independent draws. Without going into detail, note that a variety of algorithms propose generators of this type. One example, which is no longer particularly widespread, is the *Mersenne Twister* algorithm (MT) based on the Mersenne prime 19,937 [MAT 98]. It has a period of $2^{19937} - 1 \approx 10^{6600}$ (a period of the order of 10^{170} would be sufficient for the majority of simulations).

Starting with a generator of a uniform distribution over $(0,1)$, it is theoretically possible to build a sequence distributed with any cumulative function.

4.3.1. *The cumulative function inversion method*

Taking a real valued random variable, with a cumulative function $F(x)$, this method is based on the following result. Let,

$$F^{(-1)}(u) = \inf\{t,\, F(t) \geq u\} \qquad [4.4]$$

the inverse of the cumulative function F, the random variable $X = F^{(-1)}(U)$ follows a distribution with the cumulative function F if, and only if, U is a random variable which is uniformly distributed over the interval $(0,1)$.

Firstly, note that, by definition, the cumulative function of a real r.v. is the probability that it belongs to the interval $(-\infty, x]$. A cumulative function is a monotonously increasing function which may contain jumps and may be constant over certain intervals. A typical form is shown in Figure 4.1.

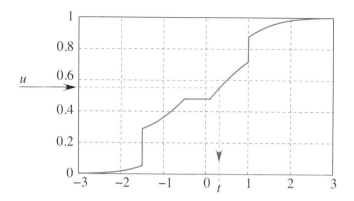

Figure 4.1. *Typical form of a cumulative function. We choose a value in a uniform manner between 0 and 1, and then deduce the realization $t = F^{(-1)}(u)$*

Take $Y = F(X)$. Using the fact that $F(x)$ is monotone, we may arrive at the following expression successively:

$$F(x) = \mathbb{P}\{X \leq x\} = \mathbb{P}\{F^{-1}(Y) \leq x\} = \mathbb{P}\{Y \leq F(x)\}$$

Finally, denoting $F(x) = y$, we have $\mathbb{P}\{Y \leq y\} = y \times \mathbb{1}(y \in (0,1))$ which is, by definition, the cumulative function of the uniform law.

EXAMPLE 4.1 (Rayleigh law generation).– A probability distribution is said to follow the Rayleigh law if its density is expressed as follows:

$$p(x) = \frac{x}{\sigma^2} e^{-x^2/2\sigma^2} \mathbb{1}(x \geq 0)$$

where $\sigma^2 > 0$. From this, we may deduce the cumulative function and its inverse.

$$u = F(x) = 1 - e^{-x^2/2\sigma^2} \Leftrightarrow F^{(-1)}(u) = \sigma\sqrt{-2\log(1-u)}$$

Write a program to create a sequence of length N following a Rayleigh law with parameter $\sigma = 1$. Compare with the theoretical probability density.

HINTS: *Type the program:*

```
# -*- coding: utf-8 -*-
"""
Created on Tue May 31 07:29:54 2016
****** rayleighsimul
```

```
@author: maurice
"""
from numpy.random import rand
from numpy import sqrt, arange, exp, log
from matplotlib import pyplot as plt
N = 100000; sigma = 1; sigma2 = sigma**2;
U=rand(N,1); X=sigma*sqrt(-2*log(1-U)); dx=arange(0,5,0.1)
rtheo = dx * exp(-(dx ** 2)/(2*sigma2))/sigma2;
plt.clf(); plt.hist(X,dx,normed='True')
plt.hold('on'); plt.plot(dx,rtheo,'k.-'); plt.hold('off')
```

■

EXERCISE 4.1 (Multinomial law).– (see p. 251) A multinomial random variable is a random variable with values in $\mathcal{A} = \{a_1, \ldots, a_P\}$ such that $\mathbb{P}\{X = a_i\} = \mu_i$, where $\mu_i \geq 0$ and $\sum_i \mu_i = 1$. The associated cumulative function is written as follows:

$$F(x) = \sum_{i=1}^{P} \mu_i \mathbb{1}(a_i \leq x)$$

and the inverse is written as follows:

$$F^{(-1)}(u) = \min\{a_j \in \mathcal{A} : \sum_{i=1}^{j} \mu_i \geq u\}$$

Write a program which creates a sequence of length N following a multinomial law with values in $\{1, 2, 3, 4, 5\}$, for which the probabilities are $0.1, 0.2, 0.3, 0.2$, and 0.2 respectively.

EXERCISE 4.2 (Homogeneous Markov chain).– (see p. 252) Consider a sequence $\{X_n\}$ where $n \in \mathbb{N}$ and which has values in the finite discrete set $\mathcal{S} = \{1, \ldots, K\}$. By definition (for more details, see the section p. 151), a sequence $\{X_n\}$ is said to be a *Markov chain* if $\mathbb{P}\{X_{n+1} = x_{n+1}|\{X_s = x_s, s \leq n\}\}$ coincides with $\mathbb{P}\{X_{n+1} = x_{n+1}|X_n = x_n\}$. The chain is said to be *homogeneous* if, in addition, the transition probabilities $\mathbb{P}\{X_{n+1} = x_{n+1}|X_n = x_n\}$ do not depend on n.

Consider a Markov chain with $K = 3$ states, the initial distribution and the transition distribution of the states are shown in Table 4.1. Write a program to create a sequence of N values from this Markov chain. Hint: exercise 4.1 may be used. Check that the estimated transition probabilities correspond to those given in Table 4.1.

$\mathbb{P}\{X_0 = 1\} = 0.5$
$\mathbb{P}\{X_0 = 2\} = 0.2$
$\mathbb{P}\{X_0 = 3\} = 0.3$

p_{ij}	$i=1$	$i=2$	$i=3$
$j=1$	0.3	0	0.7
$j=2$	0.1	0.4	0.5
$j=3$	0.4	0.2	0.4

Table 4.1. *The initial probabilities* $\mathbb{P}\{X_0 = i\}$ *and the transition probabilities* $\mathbb{P}\{X_{n+1} = j | X_n = i\}$ *of a three-state homogeneous Markov chain. We verify that, for all i,* $\sum_{j=1}^{3} \mathbb{P}\{X_{n+1} = j | X_n = i\} = 1$

4.3.2. *The variable transformation method*

4.3.2.1. *Linear transformation*

Let X be a random vector with probability density $p_X(x)$. The random vector $Y = AX + B$, where A is a square matrix taken to be invertible and B is a vector of *ad hoc* dimension, follows a probability distribution of density given by the expression [1.47] that can be written as follows:

$$p_Y(y) = \frac{p_X(A^{-1}(y - B))}{|\det\{A\}|} \qquad [4.5]$$

Thus, to simulate a random vector with a vector mean M and covariance matrix C, we must simply apply the following:

$$X = AW + M$$

where, W is a random vector of vector mean 0 and covariance matrix I and where A is a square root of C. Remember that if W is Gaussian, then X will also be Gaussian.

EXERCISE 4.3 (Linear transformation of 2D Gaussian).– (see p. 252) Write a program:

1) which uses a centered Gaussian generator of variance 1 to generate a sequence of length N of a centered Gaussian random variable, of dimension 2, with the following covariance matrix;

$$R = \begin{bmatrix} 2 & 0.95 \\ 0.95 & 0.5 \end{bmatrix}$$

2) which displays the obtained points and plots the two principal components.

4.3.2.2. *The nonlinear case*

If X follows a distribution of density $p_X(x)$ and if f is a derivable monotone function, then the random variable $Y = f(X)$ follows a distribution of density such that:

$$p_Y(y) = \frac{1}{|f'(f^{(-1)}(y))|} p_X\left(f^{(-1)}(y)\right)$$

where, $f^{(-1)}(y)$ denotes the inverse function of f and f' its derivative. This result can be extended to a bijective transformation for variables with multiple dimensions, replacing the derivative by using the determinant of the Jacobian. The example shown below concerns a function with two variables.

EXERCISE 4.4 (Box-Muller method).– (see p. 253) Using the results of example 1.1, determine an algorithm to create two centered, Gaussian random variables, of the same variance σ^2, from two independent uniform random variables in $(0, 1)$. Create a program which uses this algorithm, which is known as the *Box-Muller algorithm*.

EXERCISE 4.5 (The Cauchy distribution).– (see p. 254) A Cauchy random variable may be obtained in the following two ways:

1) consider a random variable U which is uniform over $(0, 1)$. Note that $Z = z_0 + a\tan(\pi(U - 1/2))$. Determine the distribution of Z.

The distribution of Z is a Cauchy distribution of parameters (a, z_0) and is noted to be $\mathcal{C}(z_0, a)$. It has no moment;

2) consider two centered, independent Gaussian variables X and Y of variance 1. We then construct $Z = z_0 + aY/X$ where, $a > 0$;

3) write a program which creates a sample following the Cauchy distribution of parameters $a = 0.8$ and $z_0 = 10$, using both methods. Compare the obtained histograms to the probability density of the Cauchy distribution. Plot QQplots of both sample sets.

4.3.2.3. *Sequence of correlated variables*

It may be useful to produce a sequence of correlated random variables. ARMA(P, Q) processes, see [BLA 14], are a key way. They are defined by the following recurrence equation:

$$X_n + a_1 X_{n-1} + \cdots + a_P X_{n-P} = W_n + \cdots + b_Q W_{n-Q}$$

where, W_n is a centered white noise of variance σ^2 and where all of the roots of the polynomial (as shown in the following expression)

$$A(z) = z^P + a_1 z^{P-1} + \cdots + a_P$$

are strictly within the unit circle. In theory, it is easy to determine a closed form expression of the series of covariances, but the calculation can be done recursively using the function arma2ACF provided in exercise 5.3.4.

4.3.3. *Acceptance-rejection method*

In the *acceptance-rejection method*, we use an auxiliary distribution $q(x)$ which is known to be "easy" to generate in order to construct samples with a distribution $p(x)$ that is considered as "difficult". Moreover, we consider that a value M exists such that, for any x, $Mq(x) \geq p(x)$. The acceptance-rejection algorithm consists of:

Data: $p(x)$, $q(x)$ distributions s.t. $\exists M \geq 0$ s.t. $\forall x : Mq(x) \geq p(x)$
draw X under q distribution;
emphindependently, draw U under $\mathcal{U}(0,1)$;
if $UMq(X) \leq p(X)$ **then**
 | accept $Y = X$;
else
 | reject
end

Algorithm 12: Acceptance-rejection algorithm

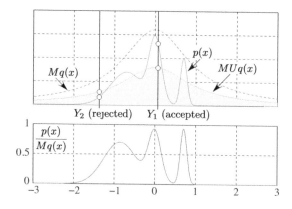

Figure 4.2. *The greater the value of M, the more samples are required to be drawn before accepting a value. The curve noted $MUq(x)$ in the figure corresponds to a draw of the r.v. U from the interval 0 to 1. For this draw, the only accepted values of Y are those for which $p(x) \geq MUq(x)$*

Let us show that Y is distributed with a probability density p.

PROOF.– Let Z be the random variable which takes a value of 0 if the draw is rejected and a value of 1 otherwise. The probability that $Z = 1$ is written as follows:

$$\mathbb{P}\{Z = 1\} = \mathbb{P}\left\{U \leq \frac{p(X)}{Mq(X)}\right\}$$

Taking account of the fact that U and X are independent, we can arrive at the following expression:

$$\mathbb{P}\{Z = 1\} = \iint_{\{(u,t):u \leq \frac{p(t)}{Mq(t)}\}} \mathbb{1}(u \in (0,1)) du \times q(t) dt$$

$$= \int_{-\infty}^{+\infty} \frac{p(t)}{Mq(t)} \times q(t) dt = \frac{1}{M}$$

To obtain this equation, we begin by integrating with respect to u then with respect to t. Now, let us calculate the conditional probability.

$$\mathbb{P}\{Y \leq x | Z = 1\} = \frac{\mathbb{P}\{Y \leq x, Z = 1\}}{\mathbb{P}\{Z = 1\}} = \frac{\mathbb{P}\left\{X \leq x, U \leq \frac{p(X)}{Mq(X)}\right\}}{\mathbb{P}\{Z = 1\}}$$

$$= M \iint_{\{t \leq x, u \leq \frac{p(t)}{Mq(t)}\}} \mathbb{1}(u \in (0,1)) du \times q(t) dt$$

$$= M \int_{-\infty}^{x} \int_{0}^{\frac{p(t)}{Mq(t)}} du \times q(t) dt$$

$$= M \int_{-\infty}^{x} \frac{p(t)}{Mq(t)} \times q(t) dt = \int_{-\infty}^{x} p(t) dt$$

Carrying out a derivation with respect to x, we see that the density of Y conditional on $Z = 1$ is expressed as follows:

$$p_{Y|Z=1}(x) = p(x)$$

which concludes the proof. ∎

This method has two main drawbacks, which are as follows:

– when M is large, the acceptation rate is low;

– it is not possible to know in advance how many draws will be necessary to obtain a series of N samples.

To illustrate, if $p(x)$ is the Gaussian distribution $\mathcal{N}(0, \sigma^2)$ and $q(x)$ is the Gaussian distribution $\mathcal{N}(0, 1)$, then M is equal to $1/\sigma$. If $\sigma = 0.01$, then $M = 100$ and hence the mean acceptance rate will be $1/100$.

4.3.4. *Sequential methods*

Sequential methods implement algorithms with value distributions which converge toward the distribution that we wish to simulate. Two of the most commonly used approaches are the *Metropolis-Hastings sampler* and the *Gibbs sampler*. These techniques, which are both based on the construction of a Markov chain, will be presented below. The acronym MCMC, *Monte-Carlo Markov Chain*, is used to refer to both samplers.

The structure of these algorithms means that they require a *burn-in period*, leading to the removal of a certain number of initial values. However, it is often difficult to evaluate this burn-in period. We expect that, asymptotically, the random variables are identically distributed and simultaneously and approximately independent. The main interest of these methods lies in the fact that the samples they produce may be used for high dimensional random vectors.

We shall begin by noting a number of properties of Markov chains.

4.3.4.1. *Markov chain*

A Markov chain is a discrete-time random process with values in $\mathcal{X}$ and such that the conditional distribution of X_n with respect to the algebra spanned by all past events $\{X_s; s < n\}$ coincides with the conditional distribution of X_n given only $X_{n-1} = x'$. The conditional probability measure of X_n knowing X_{n-1} is denoted $Q_n(dx; x')$. $Q_n(dx; x')$ is known as the transition law. A Markov chain is said to be homogeneous if the transition law $Q_n(dx; x')$ does not depend on n.

In cases where the set of possible values of X_n is finite and denoted as $\mathcal{X} = \{1, \ldots, K\}$, the transition law Q is characterized by the probabilities $q(x|x') = \mathbb{P}\{X_n = x | X_{n-1} = x'\}$ for any pair $(x, x') \in \mathcal{X} \times \mathcal{X}$. In this case, for any $x' \in \mathcal{X}$, we have:

$$\sum_{x \in \mathcal{X}} q(x|x') = 1$$

In cases where transition law Q has a density, noted $q(x|x')$, we can write that, for any $A \subset \mathcal{X}$ and for any $x' \in \mathcal{X}$:

$$\mathbb{P}\{X_n \in A | X_{n-1} = x'\} = \int_A q(x|x')dx$$

In this case, for any $x' \in \mathcal{X}$, we have:

$$\int_{\mathcal{X}} q(x|x')dx = 1$$

In what follows, unless stated otherwise, we will only use the notation associated with cases where the transition law possesses a density.

Let us determine a recurrence equation for the probability distribution of X_n. By using the Bayes' rule $p_{X_n X_{n-1}}(x, x') = p_{X_n|X_{n-1}}(x, x')p_{X_{n-1}}(x') = q(x|x')p_{X_{n-1}}(x')$, we deduce that:

$$p_{X_n}(x) = \int q(x|x')p_{X_{n-1}}(x')dx' \qquad [4.6]$$

A chain is said to be stationary if $p_{X_n}(x)$ is not dependent on n. This means that a density $p(x)$ exists such that:

$$p(x) = \int q(x|x')p(x')dx' \qquad [4.7]$$

Equation [4.7] may be seen as an equation with eigenvectors $p(x)$ for the transition kernel $q(x|x')$. The condition

$$p(x)q(x'|x) = p(x')q(x|x') \qquad [4.8]$$

on $p(x)$ is sufficient to satisfy [4.7]. Indeed, by integrating the two members of [4.8] with respect to x', we obtain [4.7]. Expression [4.8] is known as the *detailed balance equation*.

4.3.4.2. Metropolis-Hastings algorithm

We wish to carry out a draw using a distribution $p(x)$ which is known to be "difficult" to produce samples. Firstly, we select two conditional probability densities $r(x|x')$ and $a(x|x')$. The draw using $r(x|x')$ is considered to be "easy" and the conditional distribution $a(x|x')$ is chosen so that the following condition is verified:

$$p(x)a(x'|x)r(x'|x) = p(x')a(x|x')r(x|x') \qquad [4.9]$$

Let us consider the following acceptance-rejection algorithm:

Data: $p(x)$, $r(x|x')$, $a(x|x')$ distributions, X_{n-1}
Result: X_n
draw a sample X following the distribution $r(x|x_{n-1})$;
independently, draw B under Bernoulli distribution and such that
$\mathbb{P}\{B = 1\} = a(X|X_{n-1})$;
if $B = 1$ **then**
  | $X_n = X$;
else
  | $X_n = X_{n-1}$;
end

Algorithm 13: Sequential acceptance-rejection algorithm

Condition [4.9] implies that $p(x)$ verifies condition [4.8].

PROOF.– Let us denote:

$$P_R(x) = 1 - \int a(u'|x) r(u'|x) du'$$

The two steps of the algorithm mean that the conditional distribution of X_n given X_{n-1} has a density of $q(x|x') = a(x|x') r(x|x') + P_R(x') \mathbb{1}(x = x')$. We can thus write, successively, that:

$$p(x') \left(q(x|x') - a(x|x') r(x|x') \right) = p(x') \left(1 - P_R(x') \right) \mathbb{1}(x = x')$$
$$= p(x)(1 - P_R(x)) \mathbb{1}(x = x')$$
$$= p(x)(q(x'|x) - \rho(x'|x) r(x'|x))$$

Using [4.9], we obtain [4.8]. In conclusion, $p(x)$ is the stationary distribution associated with $q(x|x')$. ∎

In the Metropolis-Hastings algorithm, the acceptance/rejection law has the following specific expression:

$$a(x|x') = \min \left(\frac{p(x) r(x'|x)}{p(x') r(x|x')}, 1 \right) \qquad [4.10]$$

This supposes that $r(x|x') \neq 0$ for any pair (x, x'). Expression [4.10] must be shown to verify the condition [4.9]. One important property is that the target distribution $p(x)$ only needs to be known to within a multiplicative constant. The following two specific cases arise:

1) the chosen distribution $r(x|x')$ is symmetrical, and this can be writen as $r(x|x') = r(x'|x)$. Subsequently [4.10] leads to the following expression:

$$a(x|x') = \min\left(\frac{p(x)}{p(x')}, 1\right) \qquad [4.11]$$

2) the chosen distribution $r(x|x') = r(x)$ is independent of x'. Then [4.10] leads to the following expression:

$$a(x|x') = \min\left(\frac{p(x)r(x')}{p(x')r(x)}, 1\right) \qquad [4.12]$$

Data: $p(x), r(x|x')$ distributions
Result: $X_{0:N-1}$
Initialization: $X_0 = 0$;
for $n = 1$ **to** $N - 1$ **do**
    draw a sample X following the distribution $r(x|X_{n-1})$;
    perform $a(X|X_{n-1})$ following [4.11] or [4.12];
    if $a(X|X_{n-1}) \neq 1$ **then**
        independently, draw B under Bernoulli distribution with
        $\mathbb{P}\{B = 1\} = a(X|X_{n-1})$ **if** $B = 0$ **then**
            $X_n = X_{n-1}$;
        else
            $X_n = X$;
        end
    else
        $X_n = X$;
    end
end

Algorithm 14: Metropolis-Hastings algorithm

EXERCISE 4.6 (Metropolis-Hastings algorithm).– (see p. 255) We wish to calculate $I = \int x^2 p(x) dx$, where $p(x) \propto e^{-x^2/2\sigma^2}$. It is worth noting that $p(x)$ is given up to an unknown multiplicative factor (even if we know that this factor is $1/\sigma\sqrt{2\pi}$). For $r(x|x')$, we select a uniform distribution over the interval $[-5\sigma, 5\sigma]$, and hence $r(x_{t-1})/r(x) = 1$. Write a program implementing the Metropolis-Hastings algorithm to calculate the value of the integral I, which has a theoretical value of σ^2. You may test a few values of the burn-in period.

4.3.4.3. *Gibbs sampler*

When simulating multivariate samples, it is sometimes interesting to change single components individually. The *Gibbs sampler*, a specific case of the

Metropolis-Hastings algorithm, fulfills this function. Consider a distribution with a joint probability of $p_{X_1,\ldots,X_d}(x_1,\ldots,x_d)$; let us use the Metropolis-Hastings algorithm, where the proposition law $r(x|x')$ is the conditional distribution $p_{X_k|X_{-k}}(x_1,\ldots,x_d)$, with X_{-k} being the set of variables excluding X_k. The probability of acceptance/rejection, given by expression [4.10], is therefore equal to 1 for each pair (x, x'), indicating that the sample is always accepted.

Data: $p_{X_1,\ldots,X_d}(x_1,\ldots,x_d)$ distribution
Result: $X_{0:N-1}$
Initialization: $X_0 = 0$;
for $n = 1$ **to** $N - 1$ **do**
    for $k = 1$ **to** d **do**
        draw a sample X following the distribution $p_{X_k|X_{-k}}(x_1,\ldots,x_d)$;
        $X_{n,k} = X$;
    end
end

Algorithm 15: Gibbs sampler

EXERCISE 4.7 (Gibbs sampler).– (see p. 256) Consider a bivariate Gaussian distribution of mean $\mu = \begin{bmatrix} \mu_1 & \mu_2 \end{bmatrix}^T$ and the following covariance matrix:

$$C = \begin{bmatrix} \sigma_1^2 & \rho\sigma_1\sigma_2 \\ \rho\sigma_1\sigma_2 & \sigma_2^2 \end{bmatrix}$$

1) Determine the expression of the conditional distribution $p_{X_2|X_1}(x_1, x_2)$.

2) Using the Gibbs sampler, write a program which simulates a bivariate Gaussian distribution of mean $(0,0)$ and covariance matrix C. You may test a few values of the burn-in period.

4.3.4.4. Gibbs sampler in a Bayesian context

The Gibbs sampler is used in a Bayesian context. By using a statistical model characterized by a probability density of $p_X(x;\theta)$, we consider that $\theta \in T$ is a random vector, of which the distribution has a probability density $p_\Theta(\theta)$. Considering $p_X(x;\theta)$ as the conditional probability of X given Θ, it is possible to deduce the following conditional distribution of Θ given X:

$$p_{\Theta|X}(x,\theta) = \frac{p_X(x;\theta)p_\Theta(\theta)}{\int_T p_X(x;t)p_\Theta(t)dt} \propto p_X(x;\theta)p_\Theta(\theta) \qquad [4.13]$$

This expression can be used to generate samples of θ, given X, with the Gibbs sampler.

4.4. Variance reduction

4.4.1. *Importance sampling*

Consider the calculation of the following integral:

$$I = \int_{-\infty}^{+\infty} f(x)p(x)dx \qquad [4.14]$$

Without loss of generality, we may consider that $f(x) \geq 0$. This is possible as $f(x) = f^+(x) - f^-(x)$, where $f^+(x) = \max(f(x), 0)$ and $f^-(x) = \max(-f(x), 0)$ are both positive, and to note that, by linearity, $\int f(x)dx = \int f^+(x)dx - \int f^-(x)dx$.

Direct application of the Monte-Carlo method consists of drawing N independent samples $X_0, \ldots, X_{N-1}$ following distribution $p(x)$ and approximating the integral using the following expression:

$$\widehat{I}_1 = \frac{1}{N} \sum_{n=0}^{N-1} f(X_n) \qquad [4.15]$$

Based on the hypothesis that the N variables are independent and identically distributed, the law of large numbers ensures that $\widehat{I}_1$ converges in probability toward I. As the mean is $\mathbb{E}\left\{\widehat{I}_1\right\} = \mathbb{E}\left\{f(X_0)\right\} = I$, the estimator is unbiased. Its variance is given by the following expression:

$$\text{var}\left(\widehat{I}_1\right) = \frac{1}{N}\text{var}\left(f(X_0)\right) = \frac{1}{N}\left(\int_{-\infty}^{+\infty} f^2(x)p(x)dx - I^2\right) \qquad [4.16]$$

Now, consider a situation where instead of drawing N samples following the distribution associated to $p(x)$, the draw is carried out using an auxiliary distribution of density q, known as the *instrumental* or *proposition distribution*. The use of an auxiliary distribution may appear surprising, but, as we shall see, improves the results of the calculation. Let us return to expression [4.14], which may be rewritten:

$$I = \int_{-\infty}^{+\infty} f(x)\frac{p(x)}{q(x)}q(x)dx \qquad [4.17]$$

We may now apply the previous approach to function $f(x)p(x)/q(x)$ and consider the following estimation of I:

$$\widehat{I}_2 = \frac{1}{N} \sum_{n=0}^{N-1} f(X_n)h(X_n) \qquad [4.18]$$

where the values of function $h(x) = p(x)/q(x)$ for the values of $x = X_n$ are known as importance weights and the technique is called *importance sampling*. The law of large numbers implies that $\widehat{I}_2$ converges in probability toward I. By using the fact that the random variables $X_0, \ldots, X_{N-1}$ are independent and identically distributed, we may deduce the mean, such that

$$\mathbb{E}\left\{\widehat{I}_2\right\} = \mathbb{E}\left\{f(X_0)h(X_0)\right\} = \int_{-\infty}^{+\infty} f(x)h(x)q(x)dx = I$$

indicating that the estimator is unbiased. The variance can be written as follows:

$$\mathrm{var}\left(\widehat{I}_2\right) = \mathbb{E}\left\{\widehat{I}_2^2\right\} - I^2$$

From [4.17], the term I^2 does not depend on the choice of q. Consider the first term $\mathbb{E}\left\{\widehat{I}_2^2\right\}$. Based on the Schwarz inequality, we have the following expression:

$$\mathbb{E}\left\{\widehat{I}_2^2\right\} = \int f^2(x)\frac{p^2(x)}{q(x)}dx \times \underbrace{\int q(x)dx}_{=1} \geq \left(\int f(x)p(x)dx\right)^2$$

in which equality occurs if and only if $q(x) \propto f(x)p(x)$ and, by normalization,

$$\mu(x) = \frac{1}{\int f(u)p(u)du} \times f(x)p(x)$$

Notably, the variance of $\widehat{I}_2$ has a value of 0! However, this result is pointless, as it presumes knowledge of $\int f(u)p(u)du$, the quantity which we wish to calculate. Nevertheless, it shows that the "optimum" distribution needs to be as close as possible to the function being integrated. Specifically, if f is the indicator of the interval $(\alpha, +\infty)$, it is better to use an auxiliary law with a large number of values greater than α; this would not be the case while using a centered Gaussian distribution of variance 1 for high values of α.

The importance sampling method may be modified by using the following quantity as the estimator of the integral I:

$$\widehat{I}_3 = \frac{N^{-1}\sum_{n=0}^{N-1} f(X_n)h(X_n)}{N^{-1}\sum_{n=0}^{N-1} h(X_n)} \qquad [4.19]$$

The law of large numbers states that

$$N^{-1} \sum_{n=0}^{N-1} h(X_n) \longrightarrow \mathbb{E}\{h(X_0)\} = 1$$

using the fact that $\mathbb{E}\{h(X_0)\} = \int \frac{p(x)}{q(x)} q(x) dx = 1$. From this, it can be seen that $\widehat{I}_3$ tends in probability toward I. The interest of expression [4.19] in relation to expression [4.17] may therefore be questioned. The response to this query is that, in certain situations, distribution p is only known within an unknown multiplicative coefficient λ. Let the available expression be noted as $\widetilde{p}(x) = \lambda p(x)$. Substituting this expression into $h(x)$, we obtain $h(x) = \widetilde{p}(x)/\lambda q(x)$ and then, by applying the result to [4.19], the unknown constant λ disappears: the knowledge of this constant is therefore not required. Introducing the normalized importance weights:

$$w(X_n) = \frac{h(X_n)}{\sum_{j=0}^{N-1} h(X_j)} \qquad [4.20]$$

expression [4.19] may be rewritten as follows:

$$\widehat{I}_3 = \sum_{n=0}^{N-1} f(X_n) w(X_n) \qquad [4.21]$$

Note that, following expression [4.20], the $w(X_n)$ values are positive and their sum has a value of 1. We may therefore begin by calculating $h(X_n)$ for all values of X_n and then calculate the values of $w(X_n)$ by normalization.

The normalized importance sampling algorithm is summarized below:

Data: $p(x)$, $q(x)$ distributions, $f(x)$ fonction to be integrated
Result: $\widehat{I} \approx \int f(x) p(x) dx$
begin
　　draw $X_{0:N-1}$ under $q(x)$-distribution;
　　perform $h(X_n) = p(X_n)/q(X_n)$;
　　perform $w(X_n) = \frac{h(X_n)}{\sum_{j=0}^{N-1} h(X_j)}$;
　　perform $\widehat{I} = \sum_{n=0}^{N-1} f(X_n) w(X_n)$;
end
`/* p(x) and q(x) do not need to be normalized            */`

Algorithm 16: Normalized importance sampling calculation

EXAMPLE 4.2 (Theoretical results of the Monte-Carlo approach).– Consider a centered Gaussian random variable of variance 1. We wish to calculate the probability of this variable being greater than a given value α. This probability can be written as the following expectation:

$$I = \mathbb{P}\{X > \alpha\} = \int_\alpha^{+\infty} p(x)dx = \int_{-\infty}^{+\infty} \mathbb{1}(x > \alpha)p(x)dx \qquad [4.22]$$

where $p(x) = \dfrac{1}{\sqrt{2\pi}}e^{-x^2/2}$

1) The direct MC approach consists of drawing N independent, centered Gaussian samples of variance 1 and approximates I by the value:

$$\widehat{I}_1 = N^{-1} \sum_{n=0}^{N-1} Y_n, \text{ where } Y_n = \mathbb{1}(X_n > \alpha)$$

Applying the central limit theorem to $\widehat{I}_1$, determine a confidence interval at $100\alpha\%$ of I.

2) Consider an approach by using the importance sampling technique with the Cauchy distribution $\mathcal{C}(0,1)$ as the proposition distribution. This leads to the following approximate value:

$$\widehat{I}_2 = N^{-1} \sum_{n=0}^{N-1} Z_n,$$

where $Z_n = \mathbb{1}(X_n > \alpha)h(X_n)$ and $h(x) = \sqrt{\dfrac{\pi}{2}}e^{-x^2/2}(1+x^2)$.

Applying the central limit theorem to $\widehat{I}_1$, determine a confidence interval at $100\alpha\%$ of I.

HINTS:
The random variables Y_n have values in $\{0,1\}$ with $\mathbb{P}\{Y_n\} = \mathbb{P}\{\mathbb{1}(X_n > \alpha)\} = I$. Consequently, $\mathbb{E}\{Y_n\} = I$ and $\text{var}(Y_n) = I(1-I)$. Moreover, the independence of the variables X_n implies that the variables Y_n will also be independent. Hence, $\widehat{I}_1$ is a random variable of mean I and variance $N^{-1}\text{var}(Y_1) = N^{-1}I(1-I)$. By applying the central limit theorem to the sequence Y_n, we obtain the following expression:

1) $\qquad \sqrt{N}(\widehat{I}_1 - I) \to \mathcal{N}(0, I(1-I))$

Hence, we deduce a confidence interval of 95% of the value of I using this method, as follows:

$$I_{95\%} = \left[\widehat{I}_1 - 1.96\sqrt{\widehat{I}_1(\widehat{I}_1 - 1)/N}, \widehat{I}_1 + 1.96\sqrt{\widehat{I}_1(\widehat{I}_1 - 1)/N}\right]$$

2) $\mathbb{E}\left\{\widehat{I}_2\right\} = I$. By applying the central limit theorem to the sequence Z_n, we obtain the following expression:

$$\sqrt{N}(\widehat{I}_2 - I) \to \mathcal{N}(0, \text{var}(Z_0))$$

where,

$$\text{var}(Z_0) = \frac{\pi}{2}\int_\alpha^{+\infty} e^{-x^2}(1+x^2)^2 dx - I^2$$

which may be compared to the variance $\text{var}\left(\widehat{I}_1\right) = I(1-I)$. Numerically we can show that, for large values of α, $\widehat{I}_2$ will be better than $\widehat{I}_1$.

■

EXERCISE 4.8 (Importance sampling).– (see p. 257) We want to conduct a simulation to compare the dispersion of the estimators of I with the theoretical values. The integral to perform is $\mathbb{P}\{X > \alpha\}$ where X is Gaussian, centered, and of variance 1, as in example 4.2. Let us consider three cases. In the first case, samples are produced following the Gaussian distribution, in the second case, samples are produced following the Cauchy distribution, and in the third case, samples are produced following the Cauchy distribution, but considering that we only know that the probability density of X is proportional to $e^{-x^2/2}$.

The program will perform the theoretical and the empirical quadratic errors of the 3 estimators.

4.4.2. *Stratification*

Returning to the calculation of the integral:

$$I = \int_\mathbb{R} f(x)p_X(x)dx \qquad [4.23]$$

To reduce the variance of the estimator associated with a drawing of N i.i.d. samples, following a distribution characterized by the probability density $p_X(x)$, the stratification method involves splitting $\mathbb{R}$ into "strata" and optimizing the number of points to draw in each of these strata. As we shall see, this technique requires fulfilling the following criteria:

– the ability to calculate $\mathbb{P}\{X \in A\}$ for any segment A;
– the ability to carry out a random draw conditional on $X \in A$.

Let $A_1, \ldots, A_S$ be a partition of $\mathbb{R}$ and

$$p_s = \mathbb{P}\{X \in A_s\} = \int_{A_s} p_X(x)dx \qquad [4.24]$$

with $\sum_{s=1}^{S} p_s = 1$.

We deduce the conditional probability density of X given $X \in A_s$:

$$p_{X|X \in A_s}(x) = \frac{1}{p_s} p_X(x) \mathbb{1}_{A_s}(x) \qquad [4.25]$$

It follows that the integral I may be rewritten as follows:

$$I = \sum_{s=1}^{S} p_s \int_{\mathbb{R}} f(x) p_{X|X \in A_s}(x)dx \qquad [4.26]$$

By substituting [4.25], the following equation is obtained:

$$\int_{\mathbb{R}} f(x) p_X(x) dx = \sum_{s=1}^{S} p_s \int_{\mathbb{R}} f(x) \underbrace{\frac{1}{p_s} p_X(x) \mathbb{1}_{A_s}(x)}_{p_{X|X \in A_s}(x)} dx$$

Hence, the idea of carrying out independent draws from the S strata under conditional distributions was born. Let $X_{k_s}^{(s)}$ be the N_s variables of stratum s distributed following $p_{X|X \in A_s}(x)$. An approximate value of I may be obtained from the following expression:

$$\widehat{I}_{\text{st}} = \sum_{s=1}^{S} p_s \frac{1}{N_s} \sum_{k_s=1}^{N_s} f(X_{k_s}^{(s)})$$

From this result, we can see that $\mathbb{E}\left\{\widehat{I}_{\text{st}}\right\} = I$ and estimator $\widehat{I}_{\text{st}}$ is therefore unbiased. Let us now determine the variance. Using the hypothesis of independent drawings in the S strata, we have:

$$\text{var}\left(\widehat{I}_{\text{st}}\right) = \sum_{s=1}^{S} \frac{p_s^2}{N_s^2} \sum_{k_s}^{N_s} \text{var}\left(f(X)|X \in A_s\right) = \sum_{s=1}^{S} \frac{1}{N_s} p_s^2 \sigma_s^2 \qquad [4.27]$$

where,

$$\sigma_s^2 = \text{var}\left(f(X)|X \in A_s\right) = \int_{\mathbb{R}} (f(x) - I_s)^2 p_{X|X \in A_s}(x) dx$$

$$= \frac{1}{p_s} \int_{A_s} f^2(x) p_X(x) dx - I_s^2 \qquad [4.28]$$

It is interesting to note that, in expression [4.26], the integral

$$\int_{\mathbb{R}} f(x) p_{X|X \in A_s}(x) dx \qquad [4.29]$$

is interpreted as the conditional expectation of $f(X)$ given $Y_s = \mathbb{1}(X \in A_s)$. Following property 5 of the properties 1.10, $\text{var}\left(\mathbb{E}\left\{f(X)|Y_s\right\}\right) \leq \text{var}\left(f(X)\right)$ and therefore, $\text{var}\left(\widehat{I}\right) \leq \text{var}\left(\widehat{I}_1\right)$ where, $\widehat{I}_1$ is given by [4.15], i.e. an approximation of I by using a single stratum.

It is now possible to minimize [4.27] under the constraint $N = \sum_{s=1}^{S} N_s$. Canceling the derivative of the Lagrangian $\mathscr{L}(N_s) = \sum_{s=1}^{S} N_s^{-1} p_s^2 \sigma_s^2 + \lambda(N - \sum_{s=1}^{S} N_s)$, we obtain the following expression:

$$N_s = \left\lfloor N \frac{p_s \sigma_s}{\sum_{s=1}^{S} p_s \sigma_s} \right\rfloor$$

This solution is unusable, as it requires knowledge of σ_s and thus, of the value of the integral I. One solution (not ideal) is to take $N_s = \lfloor N p_s \rfloor$, yielding the following variance:

$$\text{var}\left(\widehat{I}_{\text{st}}\right) = \frac{1}{N} \sum_{s=1}^{S} p_s \sigma_s^2$$

To use this method, we now need to carry out a random draw following the conditional distribution $p_{X|X \in A_s}(x)$. One simple approach would be to carry out a random draw by using the distribution $p_X(x)$ and then to conserve only those values situated within the interval A_s. The problem with this approach is that the number of values to draw is random. This drawback may be overcome if (i) we use the property of inversion of the cumulative function (see section 4.3.1), which uses the uniform distribution over $(0,1)$ and if (ii) we note that for a uniform random variable U, the conditional distribution of U given that U belongs to an interval B is itself uniform, and is written as $p_{U|U \in B}(u) = \mathbb{1}(u \in B)/\ell$, where ℓ denotes the length of B.

In summary, to calculate [4.23]:

– we choose S intervals $(a_0, a_1],]a_1, a_2], \ldots,]a_{S-1}, a_S)$, where $a_0 = -\infty$ and $a_S = +\infty$;

– we calculate the S integer values,

$$N_1 = \lfloor Np_1 \rfloor,$$

$$\vdots$$

$$N_s = \left\lfloor N \sum_{j=1}^{s} p_j \right\rfloor - \left\lfloor N \sum_{j=1}^{s-1} p_j \right\rfloor,$$

$$\vdots$$

$$N_S = \left\lfloor N \sum_{j=1}^{S} p_j \right\rfloor - \left\lfloor N \sum_{j=1}^{S-1} p_j \right\rfloor$$

where, $p_s = F_X(a_s) - F_X(a_{s-1})$ and $F_X(x)$ denotes the cumulative function associated with $p_X(x)$;

– we calculate $[b_0, b_1],]b_1, b_2], \ldots,]b_{S-1}, b_S]$ where $b_s = F_X(a_s)$ with $b_0 = 0$ and $b_S = 1$;

– for each value of s, we draw N_s values $U_1^s, \ldots, U_{N_s}^s$, independently and following a uniform distribution $\mathcal{U}(b_{s-1}, b_s)/(b_s - b_{s-1})$;

– for each value of s, we calculate $X_1^{(s)} = F_X^{[-1]}(U_1^s), \ldots, X_{N_s}^{(s)} = F_X^{[-1]}(U_{N_s}^s)$;

– we calculate:

$$\widehat{I} = \sum_{s=1}^{S} p_s \frac{1}{N_s} \sum_{k_s=1}^{N_s} f(X_{k_s}^{(s)})$$

EXERCISE 4.9 (Stratification).– (see p. 258) In this exercise, we shall use the Monte-Carlo method with and without stratification to compare the calculation of the following integral:

$$I = \int_{-\infty}^{\infty} \cos(ux) \frac{1}{\sqrt{2\pi}} e^{-x^2/2} dx$$

1) By using the characteristic function $\mathbb{E}\{e^{juX}\}$, determine the analytical expression of I.

2) Write a program which compares the calculation with and without stratification.

4.4.3. *Antithetic variates*

We wish to estimate the following integral:

$$I = \int_{\mathbb{R}} f(x) p_X(x) dx \qquad [4.30]$$

The *antithetic variates method* consists of finding a pair of random variables $(X, \widetilde{X})$ such that functions $f(X)$ and $f(\widetilde{X})$ have the same expectation, the same variance, and satisfy the condition as in the following expression:

$$\text{cov}\left(f(X), f(\widetilde{X})\right) < 0 \qquad [4.31]$$

Using an i.i.d. series of random variables X_n distributed following the law $p_X(x)$, we consider the following estimator of I:

$$\widehat{I}_a = \frac{1}{N} \sum_{n=0}^{N/2-1} \left(f(X_n) + f(\widetilde{X}_n)\right) \qquad [4.32]$$

Hence,

$$\text{var}\left(\widehat{I}_a\right) = \frac{1}{N}\left(\text{var}\left(f(X)\right) + \text{cov}\left(f(X), f(\widetilde{X})\right)\right) < \frac{1}{N}\text{var}\left(f(X)\right) \qquad [4.33]$$

EXERCISE 4.10 (Antithetic variates approach).– (see p. 259) We wish to calculate the following integral:

$$I = \int_0^1 \frac{1}{1+x} dx = \log(2)$$

Integral I may be seen as the expectation of $f(X) = 1/(1+X)$ under the uniform distribution $\mathcal{U}(0,1)$. Take $\widetilde{X} = 1 - X$. Hence, if $X \sim \mathcal{U}(0,1)$, then $\widetilde{X} \sim \mathcal{U}(0,1)$.

Let $\widehat{I}$ and $\widehat{I}_a$ be the estimators given by [4.14] and [4.32] respectively.

1) Determine the values of $\text{var}(f(X))$; $\text{cov}\left(f(X), f(\widetilde{X})\right)$; $\text{var}\left(\widehat{I}\right)$; and $\text{var}\left(\widehat{I}_a\right)$ for $N = 100$.

2) Write a program to compare the direct calculation with the antithetic variate method.

5

Hints and Solutions

5.1. Useful Maths

5.1.1.– (Module and phase joint law of a 2D Gaussian r.v.) (see p. 20) As this substitution is bijective and its Jacobian is equal to r, which is positive, the joint distribution of the pair (R, Θ) has a density of:

$$p_{R\Theta}(r, \theta) = r\, p_{XY}(r\cos(\theta), r\sin(\theta))\mathbb{1}(r \geq 0)\mathbb{1}(\theta \in (0, 2\pi))$$

$$= \frac{r}{2\pi\sigma^2} e^{-r^2/2\sigma^2} \mathbb{1}(r \geq 0)\mathbb{1}(\theta \in (0, 2\pi))$$

Following property [1.3] we derive from $p_{R\Theta}(r,\theta) = g(r)h(\theta)$, where $g(r) = \frac{r}{\sigma^2} e^{-r^2/2\sigma^2}\mathbb{1}(r \geq 0)$ and $h(\theta) = (2\pi)^{-1}\mathbb{1}(\theta \in (0, 2\pi))$, that the random variables R and Θ are independent. Therefore, Θ is uniform on $(0, 2\pi)$ and R has a Rayleigh distribution.

5.1.2.– (δ-method) (see p. 21) We shall use the hypothesis $\mathrm{cov}(X_0, X_1) = \sigma^2 I_2$. The Jacobian, here noted ∂g, of $g : (X_0, X_0) \to (R, \theta)$ is deduced from the Jacobian ∂h of $h : (R, \theta) \to (X_0, X_1)$ following:

$$\partial h = \begin{bmatrix} \cos(\theta) & -R\sin(\theta) \\ \sin(\theta) & R\cos(\theta) \end{bmatrix} \Rightarrow \partial g = \partial h^{-1} = \frac{1}{R}\begin{bmatrix} R\cos(\theta) & R\sin(\theta) \\ -\sin(\theta) & \cos(\theta) \end{bmatrix}$$

We have:

$$\mathrm{cov}\left(\begin{bmatrix} R \\ \theta \end{bmatrix}\right) = \begin{bmatrix} \mathrm{cov}(R,R) & \mathrm{cov}(R,\theta) \\ \mathrm{cov}(R,\theta) & \mathrm{cov}(\theta,\theta) \end{bmatrix} = \begin{bmatrix} \mathrm{var}(R) & \mathrm{cov}(R,\theta) \\ \mathrm{cov}(R,\theta) & \mathrm{cov}(\theta,\theta) \end{bmatrix}$$

and, from [1.49]:

$$\operatorname{cov}\left(\begin{bmatrix}R\end{bmatrix}\right) \approx \partial g \operatorname{cov}(X_0, X_1) \partial^T g = \sigma^2 I_2$$

We deduce:

$$\operatorname{var}(R) = \begin{bmatrix}1 & 0\end{bmatrix} \sigma^2 \begin{bmatrix}1 \\ 0\end{bmatrix} = \sigma^2 \qquad [5.1]$$

This result is therefore correct on the condition that $(\mu_0^2 + \mu_1^2)/\sigma^2$ is large. This result may appear weak; note, however, that the δ-method allows us to calculate an approximation even in cases where $\operatorname{cov}(X_0, X_1)$ is different from $\sigma^2 I_2$.

Let $\nu = \sqrt{\mu_0^2 + \mu_1^2}$. Program deltaMethodRice.m estimates the variance of R by carrying out a simulation using 100000 draws. If $\nu/\sigma > 4$, then the approximation is acceptable. Note that the function g under consideration is not differentiable at point $(0,0)$, associated with the case where $\mu_0 = \mu_1 = 0$. Type the program:

```
# -*- coding: utf-8 -*-
"""
Created on Sun May 29 07:47:37 2016
****** deltaMethodRice
@author: maurice
"""
from numpy.random import rand, randn
from numpy import sin, cos,sqrt,pi,std,sum
Lruns = 10000;
nu = 8.0; sigma = 2.0; a = 2*rand()*pi;
mu = [nu*cos(a),nu*sin(a)];
X = mu + sigma * randn(Lruns,2);
R = sqrt(sum(X ** 2, axis=1));
print('true value = %4.2f, simulation = %4.2f'%(sigma,std(R)))
```

5.1.3.– (Asymptotic confidence interval from the CLT) (see p. 23)

1) From the hypotheses, we deduce that, for any k, $\mathbb{E}\{X_k\} = p$ and $\operatorname{var}(X_k) = p(1-p)$. According to the central limit theorem 1.9, when N tends toward infinity, we have:

$$\sqrt{N}(\widehat{p} - p) \xrightarrow{d} \mathcal{N}(0, p(1-p))$$

2) From this, we deduce:

$$\mathbb{P}\left\{-\epsilon \leq \sqrt{N}(\widehat{p} - p) \leq \epsilon\right\} \approx \int_{-\epsilon}^{\epsilon} \frac{1}{\sqrt{2\pi p(1-p)}} e^{-\frac{u^2}{2p(1-p)}} du$$

Taking $w = u/\sqrt{p(1-p)}$ and $\delta = \epsilon/\sqrt{p(1-p)}$, we obtain:

$$\mathbb{P}\left\{-\delta\sqrt{p(1-p)} \leq \sqrt{N}(\widehat{p}-p) \leq \delta\sqrt{p(1-p)}\right\}$$

$$\approx \int_{-\delta}^{\delta} \frac{1}{\sqrt{2\pi}} e^{-w^2/2} dw$$

3) Solving the double inequality in terms of p, we deduce the confidence interval at $100\,\alpha\%$:

$$\mathrm{CI}_{100\alpha\%} = \left(\frac{\widehat{p}+\frac{\delta^2}{2N}-\sqrt{\Delta}}{1+\frac{\delta^2}{N}}, \frac{\widehat{p}+\frac{\delta^2}{2N}+\sqrt{\Delta}}{1+\frac{\delta^2}{N}}\right) \quad [5.2]$$

where $\Delta = \widehat{p}(1-\widehat{p})\frac{\delta^2}{N} + \frac{\delta^4}{4N^2}$ and where δ is linked to the value α by:

$$\int_{-\delta}^{\delta} \frac{1}{\sqrt{2\pi}} e^{-w^2/2} dw = \alpha$$

Typically, for $\alpha = 0.95$, $\delta = 1.96$.

4) Type the program:

```
# -*- coding: utf-8 -*-
"""
Created on Sun May 29 08:26:42 2016
******** ICpercent
@author: maurice
"""
from numpy import zeros, sum, mean, sqrt
from scipy.stats import norm
from numpy.random import rand
Lruns = 1000; N = 300; alpha = 0.9;
delta = norm.isf(1.0-(1.0-alpha)/2.0);
p = 0.2; CI = zeros([2,Lruns]);
for ir in range(Lruns):
    X = rand(N) < p;
    hatp = mean(X);
    Delta = hatp*(1-hatp)*(delta**2)/N+(delta**4)/4/N/N;
    CI[0,ir] = (hatp+(delta**2)/2/N-sqrt(Delta))/(1+delta**2/N);
    CI[1,ir] = (hatp+(delta**2)/2/N+sqrt(Delta))/(1+delta**2/N);
# percent of the true value inside the confidence interval
aux = (CI[0,:]<p) & (CI[1,:]>p)
percentinCI=100.0*sum(aux)/Lruns;
print('*****Percent in the confidence interval:%3.2f'%percentinCI)
```

5.2. Statistical inferences

5.2.1.– **(Iris classification)** (see p. 40) First, type the following module:

```
# -*- coding: utf-8 -*-
"""
Created on Sun Aug 14 16:28:28 2016
****** pcaldatoolbox
@author: maurice
"""
from numpy import zeros, mean, size, unique, sum
from numpy import matrix as mat
from scipy.linalg import eig
#==============================
def LDA(X,y,k):
    """
    # SYNOPSIS: LDA(X,y,k)
    # inputs:
    #    X: explanatory variables, array N x d
    #    y: class indexes, array N x 1
    #    k: integer (less than d)
    # outputs:
    #    Xtilde: (N x k) matrix of reduced variables
    #    V: d x k array
    """
    N = size(X,0);
    d = size(X,1)
    labely = unique(y);   g = len(labely);
    meanell = zeros([g,d]); Nell = zeros(g)
    for ig in range(g):
        meanell[ig,:] = mean(X[y==labely[ig],:],0)
        Nell[ig] = sum(y==labely[ig])
    Xc = zeros([N,d])
    for ig in range(g):
        Xc[y==labely[ig],:]=X[y==labely[ig],:]-meanell[ig,:]
    RI = mat(zeros([d,d]));
    meanG = mat(zeros(d));
    for ig in range(g):
        Xc_ig = mat(Xc[y==labely[ig],:])
        mataux = Xc_ig.T*Xc_ig
        X_ig = mat(X[y==labely[ig],:])
        meanG = meanG + sum(X_ig,0)
        RI = RI+mataux;
    RI = RI/N
```

```
        meanG = meanG/N;
        RE = mat(zeros([d,d]));
        for ig in range(g):
            vaux = mat(meanell[ig,:]-meanG);
            RE = RE+Nell[ig]*(vaux.T*vaux);
        RE = RE/N;
        #=====   [V,D] = eig(RE,RI) solves RE*V = lambda*RI*V
        generaleigdecompo = eig(RE,RI);
        eigvectors = generaleigdecompo[1]
        eigvalues = generaleigdecompo[0].real
        eigvectorssort = eigvectors[:,eigvalues.argsort()]
        V = mat(eigvectorssort[:,d-k:d]);
        Xc = mat(Xc)
        Xtilde = X * V;
        return Xtilde, V
#================================
def PCA(X,k):
    """
    # SYNOPSIS: LDA(X,k)
    # inputs:
    #    X: observations, array N x d
    #    k: integer (less than d)
    # outputs:
    #    Xtilde: (N x k) matrix of reduced variables
    #    V: d x k array
    """
    Xmat = mat(X);
    N = size(Xmat,0)
    d = size(Xmat,1)
    meanX = mean(Xmat,0)
    Xc = mat(zeros([N,d]))
    for indd in range(d):
        Xc[:,indd] = Xmat[:,indd] - meanX[0,indd]
    covX = Xc.transpose()*Xc/N
    eigdecompo = eig(covX)
    eigvectors = eigdecompo[1]
    eigvalues = eigdecompo[0]
    eigvectorssort = eigvectors[:,eigvalues.argsort()]
    V = mat(eigvectorssort[:,range(d-k,d)])
    Xtilde = Xc * V
    return Xtilde, V
```

Type and run the below program. We see that the class Setosa is linearly separable from the other two. Note that in the LDA, we use the iris categories, while in the PCA, the reduction is applied in "blind" using the data, all together.

```
# -*- coding: utf-8 -*-
"""
Created on Mon Jul  4 16:05:14 2016
****** irispcalda
@author: maurice
"""
from sklearn import datasets
from matplotlib import pyplot as plt
from pcaldatoolbox import LDA, PCA
from numpy import unique
#============ main program
iris = datasets.load_iris(); #print(datasets.load_iris().DESCR)
X = iris.data; y = iris.target; t_names = iris.target_names
k = 2; labely = unique(y); g = len(labely)
XPCA, VPCA = PCA(X,k); XLDA, VLDA = LDA(X,y,k);
plt.clf(); plt.subplot(121)
for col, ig, target_name in zip("rgb", range(g), t_names):
    plt.scatter(XPCA[y==labely[ig],0], XPCA[y==labely[ig],1], \
        c=col, label=target_name)
    plt.hold('on')
plt.hold('off'); plt.legend(); plt.title('PCA on IRIS dataset')
plt.subplot(122)
for col, ig, target_name in zip("rgb", range(g), t_names):
    plt.scatter(XLDA[y==labely[ig],0], XLDA[y==labely[ig],1], \
        c=col, label=target_name)
    plt.hold('on')
plt.hold('off'); plt.legend(); plt.title('LDA on IRIS dataset');
plt.show()
```

5.2.2.– **(Empirical ROC curve and AUCl)** (see p. 50)

1) The significance level α represents the probability that, under H_0, the statistic $\Phi(X)$ will be higher than the threshold ζ. Consequently, to estimate α for the threshold ζ, we need to count how many values of $\Phi(x)$ are greater than ζ in the database H_0. The same approach should be taken for the power; to obtain the ROC curve, we alter ζ across a range dependent on the minimum and maximum values of $\Phi(X)$. One simple approach is to rank the set of values of $\Phi(X)$ in decreasing order, and then

calculate the cumulative sum of the indicators associated with the two hypotheses. This is written as:

$$\begin{cases} \widehat{\alpha}_j = \dfrac{1}{N_0} \sum_{i_0=0}^{N_0-1} \mathbb{1}(\Phi_{0,i_0} > T_j) \\ \widehat{\beta}_j = \dfrac{1}{N_1} \sum_{i_1=0}^{N_1-1} \mathbb{1}(\Phi_{1,i_1} > T_j) \end{cases}$$

where T_j is the sequence of values ranked in decreasing order of the set $\{T_{0,i_0}\} \cup \{T_{1,i_1}\}$, where $j = 1$ to $N_0 + N_1$.

2) Type the program:

```
# -*- coding: utf-8 -*-
"""
Created on Mon Jun  6 22:22:25 2016
****** roccurve2gaussians
@author: maurice
"""
from numpy.random import randn
from numpy import zeros, cumsum, sqrt, mean, linspace, concatenate
from scipy.stats import norm
from matplotlib import pyplot as plt
N0 = 1000; N1 = 1200; Nt = N0+N1; n = 10; m0 = 0; m1 = 0.5;
#===== data bases H0 and H1
X0 = m0+randn(N0,n); X1 = m1+randn(N1,n);
#===== statistics Phi(X)
Phi0 = mean(X0,1); Phi1 = mean(X1,1);
#===== Experimental ROC curve estimate
c0 = zeros(Nt); c1 = zeros(Nt);
idx = concatenate((Phi0,Phi1)).argsort();
c0[idx<=N0] = 1; hatalpha = cumsum(c0)/N0;
c1[idx>N0] = 1; hatbeta = cumsum(c1)/N1;
hatalpha = concatenate((zeros(1),hatalpha));
hatbeta = concatenate((zeros(1),hatbeta));
zeta = linspace(0,10,200);
#===== theoretical ROC curve
alpha = 1-norm.cdf(zeta,m0,1/sqrt(n));
beta = 1-norm.cdf(zeta,m1,1/sqrt(n));
plt.clf(); plt.plot(1.0-hatalpha,1.0-hatbeta)
plt.hold('on'); plt.plot(alpha,beta,'.--');
plt.hold('off'); plt.xlim([0,1]); plt.ylim([0,1])
#===== EAUC
W = 0;
for i0 in range(N0):
```

```
    for i1 in range(N1):
        W = W+(Phi1[i1]>Phi0[i0])+0.5*(Phi1[i1]==Phi0[i0]);
eauc = W/(N0*N1);
print('Experimental AUC = %5.2f'%eauc);
```

5.2.3.– (**Student distribution for** H_0) (see p. 53) In the following program, the variance and the mean have been chosen at random, and we see that they have no effect on the distribution of $V(X)$.

Type the following program:

```
# -*- coding: utf-8 -*-
"""
Created on Tue Jun  7 05:33:10 2016
****** studentlawmdiffm0
@author: maurice
"""
from numpy.random import rand, randn
from numpy import mean, ones, sqrt, dot, sum
from scipy.stats import t
from matplotlib import pyplot as plt
sigma = rand()*100; L = 10000; n = 100; m0 = 10*randn();
X = randn(n,L)+m0; mhat = mean(X,axis=0)
num = (mhat-m0)*sqrt(n);
Xc = X - dot(ones([n,1]),mhat.reshape(1,L))
denum = sqrt(sum(Xc **2,axis=0)/(n-1));
V = num / denum;
plt.clf(); aux = plt.hist(V,bins=100,normed='True')
ttheo = t.pdf(aux[1],n-2);
plt.hold('on'); plt.plot(aux[1], ttheo,'.-r')
plt.hold('off'); plt.show()
```

5.2.4.– (**Unilateral mean testing**) (see p. 54) The log-likelihood is written as:

$$\mathcal{L}(\theta) = -\frac{n}{2}\log(2\pi) - \frac{n}{2}\log(\sigma^2) - \frac{1}{2\sigma^2}\sum_{k=1}^{n}(X_k - m)^2$$

First, maximizing with respect to σ^2, we obtain:

$$\widetilde{\mathcal{L}}(m) = -\frac{n}{2}\log(2\pi) - \frac{n}{2}\log\sum_{k=1}^{n}(X_k - m)^2$$

The maximum on the subset H_0 is obtained for $m = n^{-1}\sum_{k=1}^{n} X_k$ on the condition that $n^{-1}\sum_{k=1}^{n} X_k \geq m_0$, otherwise it is obtained for m_0. The maximum for H_0 is therefore written as:

$$-\frac{n}{2}\log(2\pi) - \frac{n}{2}\log \sum_{k=1}^{n}(X_k - \widehat{m})^2 \mathbb{1}(\widehat{m} \geq m_0) - \frac{n}{2}\log \sum_{k=1}^{n}(X_k - m_0)^2 \mathbb{1}(\widehat{m} < m_0)$$

The maximum on the full set Θ is expressed $-\frac{n}{2}\log(2\pi) - \frac{n}{2}\log \sum_{k=1}^{n}(X_k - \widehat{m})^2$.

The log-GLRT, noted A, is written as:

$$A = -\frac{n}{2}\log \sum_{k=1}^{n}(X_k - \widehat{m})^2 \left(\mathbb{1}(\widehat{m} \geq m_0) + \mathbb{1}(\widehat{m} < m_0)\right)$$

$$+\frac{n}{2}\log \sum_{k=1}^{n}(X_k - \widehat{m})^2 \mathbb{1}(\widehat{m} \geq m_0)$$

$$+\frac{n}{2}\log \sum_{k=1}^{n}(X_k - m_0)^2 \mathbb{1}(\widehat{m} < m_0)$$

$$A = -\frac{n}{2}\left(\log \frac{\sum_{k=1}^{n}(X_k - \widehat{m})^2}{\sum_{k=1}^{n}(X_k - m_0)^2}\right) \mathbb{1}(\widehat{m} < m_0)$$

$$= -\frac{n}{2}\left(\log \frac{\sum_{k=1}^{n}(X_k - \widehat{m})^2}{\sum_{k=1}^{n}(X_k - \widehat{m} + (\widehat{m} - m_0))^2}\right) \mathbb{1}(\widehat{m} < m_0)$$

$$= -\frac{n}{2}\left(\log \frac{\sum_{k=1}^{n}(X_k - \widehat{m})^2}{\sum_{k=1}^{n}(X_k - \widehat{m})^2 + n(\widehat{m} - m_0)^2}\right) \mathbb{1}(\widehat{m} < m_0)$$

$$A = \frac{n}{2}\log(1 + T(X)) \, \mathbb{1}(\widehat{m} < m_0) \qquad [5.3]$$

taking:

$$T(X) = \frac{(\widehat{m} - m_0)^2}{n^{-1}\sum_{k=1}^{n}(X_k - \widehat{m})^2}$$

Taking the exponential of the two members with $n \neq 0$, we obtain:

$$\text{GLRT}^{\frac{2}{n}} = (1 + T(X)) \, \mathbb{1}(\widehat{m} < m_0) + \mathbb{1}(\widehat{m} \geq m_0) = T(X) \, \mathbb{1}(\widehat{m} < m_0)$$

and the test:

$$V(X) = T(X)\mathbb{1}(\widehat{m} < m_0) \underset{H_0}{\overset{H_1}{\gtrless}} \eta$$

If $\eta < 0$, we always decide in favor of H_1 and the significance level is $\alpha = 1$. We shall therefore use $\eta > 0$. Let:

$$U(X) = \frac{\sqrt{n}(m_0 - \widehat{m})}{(n-1)^{-1/2}\sqrt{\sum_{k=1}^{n}(X_k - \widehat{m})^2}}$$

We then verify that V is an increasing monotone function of U. Comparing V to a threshold is therefore equivalent to comparing U to a threshold. We know that, for H_0, the statistic $U(X)$ follows a Student distribution with $(n-1)$ degrees of freedom. The test is therefore written as:

$$\frac{\sqrt{n}(m_0 - \widehat{m})}{(n-1)^{-1/2}\sqrt{\sum_{k=1}^{n}(X_k - \widehat{m})^2}} \underset{H_0}{\overset{H_1}{\gtrless}} T_{n-1}^{[-1]}(1-\alpha)$$

5.2.5.– (Mean equality test) (see p. 56)

1) The model is constituted by the two independent distributions { i.i.d. $\mathcal{N}(n_0; m_0, \sigma^2)$ } and { i.i.d. $\mathcal{N}(n_1; m_1, \sigma^2)$ }. We let $\theta = (m_0, m_1, \sigma) \in \Theta = \mathbb{R} \times \mathbb{R} \times \mathbb{R}^+$. Hypothesis $H_0 = \{\theta \in \Theta : \text{s.t. } m_0 = m_1\}$.

2) Let $n = n_0 + n_1$. The maximization of the probability density for Θ leads us to take $\widehat{m}_0 = n_0^{-1}\sum_{k=0}^{n_0-1} X_{0,k}$, $\widehat{m}_1 = n_1^{-1}\sum_{k=0}^{n_1-1} X_{1,k}$ and $\widehat{\sigma}_0^2 = n^{-1}\sum_{k=0}^{n_0-1}(X_{0,k} - \widehat{m}_0)^2 + n^{-1}\sum_{k=0}^{n_0}(X_{1,k} - \widehat{m}_1)^2$.

Maximization of the probability density for H_0 leads us to take $\widetilde{m} = n^{-1}\sum_{k=0}^{n_0-1} X_{0,k} + n^{-1}\sum_{k=0}^{n_1-1} X_{1,k}$ and $\widetilde{\sigma}^2 = n^{-1}\sum_{k=0}^{n_0-1}(X_{0,k} - \widetilde{m})^2 + n^{-1}\sum_{k=0}^{n_1-1}(X_{1,k} - \widetilde{m})^2$. From this, we deduce:

$$\widetilde{\sigma}^2 = \widehat{\sigma}_0^2 + \frac{n_0}{n}(\widetilde{m} - \widehat{m}_0)^2 + \frac{n_1}{n}(\widetilde{m} - \widehat{m}_1)^2$$

and the GLRT:

$$\Lambda(X) \propto \frac{\widetilde{\sigma}^2}{\widehat{\sigma}_0^2} = 1 + \frac{\frac{n_0}{n}(\widetilde{m} - \widehat{m}_0)^2 + \frac{n_1}{n}(\widetilde{m} - \widehat{m}_1)^2}{n^{-1}\sum_{k=0}^{n_0-1}(X_{0,k} - \widehat{m}_0)^2 + n^{-1}\sum_{k=0}^{n_1-1}(X_{1,k} - \widehat{m}_1)^2}$$

Comparing the GLRT to a threshold can be shown to be equivalent to comparing the following statistic $W(X)$ to a threshold:

$$W(X) = \frac{\sqrt{n_H}\,|\widehat{m}_0 - \widehat{m}_1|}{S/\sqrt{n-2}}$$

where $n_H^{-1} = n_0^{-1} + n_1^{-1}$ and $S^2 = \sum_{k=0}^{n_0-1}(X_{0,k} - \widehat{m}_0)^2 + \sum_{k=0}^{n_1-1}(X_{1,k} - \widehat{m}_1)^2$.

3) Based on property 2.3 (section 2.6.3), we deduce that $\widehat{m}_0$, $\widehat{m}_1$, $e_0 = X_0 - \widehat{m}_0$ and $e_1 = X_1 - \widehat{m}_1$ are jointly independent and Gaussian. Hence, for H_0:

$$\frac{U}{\sigma} = \frac{\sqrt{n_H}(\widehat{m}_0 - \widehat{m}_1)}{\sigma} \sim \mathcal{N}(0,1)$$

We also deduce that $S^2/\sigma^2 \sim \chi^2_{n-2}$ and that S^2 is independent of U. Hence, in accordance with [1.61]:

$$V(X) = \frac{U(X)/\sigma}{S(X)/\sigma\sqrt{n-2}} \sim \frac{\mathcal{N}(0,1)}{\sqrt{\chi^2_{n-2}/(n-2)}} = T_{n-2}$$

where T_{n-2} is a Student variable with $(n-2)$ degrees of freedom. This distribution is symmetrical around the value 0. The test therefore takes the form:

$$W(X) \underset{H_0}{\overset{H_1}{\gtrless}} \eta$$

where $\eta = T_{n-2}^{[-1]}(1 - \alpha/2)$ and where α is the confidence level.

4) The p-value is written as:

$$p\text{-value} = 2\int_{T(X_1,X_2)}^{+\infty} T_{n-2}(t)dt$$

5) The below program provides a p-value of 0.63, so the hypothesis of mean equality is accepted.

```
# -*- coding: utf-8 -*-
"""
Created on Wed Jun  8 23:20:30 2016
****** Ttest
@author: maurice
"""
from numpy import array, mean, sqrt
from scipy.stats import t
```

```
data = array([[1, 1, 1, 1, 2, 2, 2],\
        [51.0,53.3,55.6,51.0,55.5,53.0,52.1]]);
d1 = data[1,data[0,:]==1]; d2 = data[1,data[0,:]==2];
N1=len(d1); N2=len(d2); N=N1+N2;
m1=mean(d1); m1c = d1-m1; m2=mean(d2); m2c = d2-m2;
NH = 1.0/(1.0/N1+1.0/N2); U   = abs(m1-m2)* sqrt(NH);
S2 = sum(m1c*m1c)+sum(m2c*m2c); W   = U/sqrt(S2/(N-2));
pvalue = 2*(1-t.cdf(W,N-2));
if pvalue<0.05:
    dec='H0 false'
else:
    dec='H0 true'
print('\tp-value of H0={m1=m2} = %4.2f ==> %s\n'%(pvalue,dec))
```

Program studentlawdiffm0m1.py shows by simulation that W follows a Student distribution with $n - 2$ degrees of freedom.

```
# -*- coding: utf-8 -*-
"""
Created on Tue Jun  7 05:22:19 2016
****** studentlawdiffm0m1
@author: maurice
"""
from numpy.random import rand, randn
from numpy import mean, ones, sqrt, dot, sum
from scipy.stats import t
from matplotlib import pyplot as plt
sigma = rand()*100; L = 10000;
n0 = 100; n1 = 150; n = n0+n1;
nh = 1.0 /(1.0/n0+1.0/n1); m0 = 10*randn();
X0 = randn(n0,L)+m0; X1 = randn(n1,L)+m0;
m0 = mean(X0,0); m1 = mean(X1,0);
Xc1 = X0 - dot(ones([n0,1]),m0.reshape(1,L));
Xc2 = X1 - dot(ones([n1,1]),m1.reshape(1,L));
S2 = sum(Xc1 **2,axis=0)+sum(Xc2 **2,axis=0);
U = (m0-m1)*sqrt(nh); W = U / sqrt(S2/(n-1));
plt.clf(); aux = plt.hist(W, bins=100, normed='True');
ttheo = t.pdf(aux[1],n-2);
plt.hold('on'); plt.plot(aux[1], ttheo,'.-r')
plt.hold('off'); plt.show()
```

5.2.6.– **(CUSUM algorithm)** (see p. 56)

1) The statistical model is a family of probability distributions dependent on the parameter m with values in $\{0,\ldots,n-1\}$ and a log-density which is written as:

$$\ell(x;m) = \left\{\left\{1.6 \begin{array}{l} \sum_{k=0}^{m-1} \log p(x_k;\mu_0) + \sum_{k=m}^{n-1} \log p(x_k;\mu_1) \text{ if } m \in \{0,\ldots,n-2\} \\ \sum_{k=0}^{n-1} \log p(x_k;\mu_0) \qquad\qquad\qquad\qquad\quad \text{if } m = n-1 \end{array}\right.\right.$$

Using this notation, the hypothesis to test is $H_0 = \{n-1\}$, meaning there is no change.

2) For the sample of length n, the test function of the GLRT is written as:

$$T_n(X) = \max_{0 \le m \le n-2} \ell(X;m) - \ell(X;n-1)$$

$$= \sum_{k=0}^{n-1} s_k - \min_{1 \le m \le n} \sum_{k=0}^{m-1} s_k \qquad [5.4]$$

taking $s_k = \log p(x_k;\mu_1)/p(x_k;\mu_0)$. Hence, $T_n(X) \ge 0$.

3) Presuming that we know the test function for a sample of size $n-1$ and that we are observing a new value s_n, $C(n)$ is either greater or less than $\min_{0 \le m \le n-2} \sum_{k=0}^{m-1} s_k$. If $C(n) > \min_{0 \le m \le n-2} \sum_{k=0}^{m-1} s_k$, then $\min_{0 \le m \le n-1} \sum_{k=0}^{m-1} s_k = \min_{0 \le m \le n-2} \sum_{k=0}^{m-1} s_k$ and $T_n = C(n) - \min_{0 \le m \le n-2} \sum_{k=0}^{m-1} s_k = T_{n-1} + s_n$. If $C(n) \le \min_{0 \le m \le n-2} \sum_{k=0}^{m-1} s_k$, then $T_n = C(n) - C(n) = 0$. As $T_n \ge 0$, we deduce that:

$$T_n = \max\{T_{n-1} + s_n, 0\} \qquad [5.5]$$

The following program verifies that the direct formula [5.4] gives the same values as the recursive expression of the CUSUM, i.e. [5.5].

```
# -*- coding: utf-8 -*-
"""
Created on Wed Jun  1 12:22:50 2016
****** CUSUMrecursiveformula
@author: maurice
"""
from numpy.random import randn
from numpy import cumsum, max, zeros, min
from matplotlib import pyplot as plt
n=160; s=randn(n); C=cumsum(s);
```

```
plt.clf()
Tdirect = zeros(n); Trecursive = zeros(n);
for ii in range(1,n):
    Tdirect[ii]=C[ii]-min(C[0:ii+1]);
    Trecursive[ii]=max([Trecursive[ii-1]+s[ii],0.0]);
plt.plot(Tdirect,'.-')
plt.hold('on'); plt.plot(Trecursive,'.-')
plt.hold('off'); plt.show()
```

4) The following program implements the CUSUM test for a change in the means of two Gaussians with the same variables. The test works better as the difference between the means increases in relation to the standard deviation. We can also verify that the shape of the CUSUM does not depend on the sense of the inequality between μ_0 and μ_1.

```
# -*- coding: utf-8 -*-
"""
Created on Wed Jun  1 13:07:23 2016
****** CUSUMtest
@author: maurice
"""
from numpy.random import randn
from numpy import max, zeros, sum
from matplotlib import pyplot as plt
n = 120; m = 32; mu0lessthanmu1 = 0;
if mu0lessthanmu1:
    mu0 = 1.0; mu1 = 8.0;
else:
    mu1 = 1.0; mu0 = 8.0;
x=zeros(n);T=zeros(n); x[0:m]=randn(m)+mu0; x[m:n]=randn(n-m)+mu1;
for nn in range(1,n):
    sn=(x[nn]-mu0)**2-(x[nn]-mu1)**2;
    T[nn]=max([T[nn-1]+sn,0]);
hatm = sum(T==0)
plt.clf(); plt.subplot(211); plt.plot(x)
plt.subplot(212); plt.plot(T,'.-'); plt.hold('on')
plt.plot(hatm,T[hatm],'or'); plt.hold('off'); plt.show()
```

5.2.7.– **(Proof of** [2.37]) (see p. 58)

1) $\mathbb{E}\{N_j\} = Np_j$, using the fact that

$$\mathbb{E}\{\mathbb{1}(X_k \in \Delta_j)\} = \mathbb{P}\{X_k \in \Delta_j\}$$

In the same way, $\mathbb{E}\{N_j N_m\} = N p_j \delta(j,m) + N(N-1) p_j p_m$. Therefore:

$$C = D - PP^T$$

where $D = \mathrm{diag}(P)$.

2) We have:

$$\begin{aligned} C &= D - D^{1/2} D^{-1/2} PP^T D^{-1/2} D^{1/2} \\ &= D - D^{1/2} VV^T D^{1/2} \\ &= D^{1/2}(I - VV^T) D^{1/2} \end{aligned}$$

where $V = D^{-1/2} P = \begin{bmatrix} \sqrt{p_0} & \cdots & \sqrt{p_{g-1}} \end{bmatrix}^T$. Thus:

$$\Gamma = D^{-1/2} C D^{-1/2} = I - VV^T$$

We note that $V^T V = 1$ and that VV^T is the projector onto V which is of rank 1. Γ is therefore a projector of rank $(g-1)$. A unit matrix U therefore exists, such that:

$$\Gamma = U \begin{bmatrix} I_{g-1} & 0 \\ 0 & 0 \end{bmatrix} U^T \qquad [5.6]$$

3) Let:

$$\widehat{P} = \begin{bmatrix} \dfrac{N_0}{N} & \cdots & \dfrac{N_{g-1}}{N} \end{bmatrix}^T$$

The central limit theorem 1.9 states that the random variable

$$Y = \sqrt{N}\left(\widehat{P} - P\right) \xrightarrow{d} \mathcal{N}(0, C)$$

converges in distribution toward a Gaussian of mean vector 0 and covariance matrix C. The random vector $Z = D^{-1/2} Y$ therefore converges in distribution following:

$$Z = D^{-1/2} Y \xrightarrow{d} \mathcal{N}(0, \Gamma)$$

Using equation [5.6], $Z^T Z$ appears asymptotically as the sum of the squares of $(g-1)$ Gaussian, independent, centered random variables with variance 1. $Z^T Z$ therefore converges in distribution toward a variable of χ^2 with $(g-1)$ degrees of freedom. This demonstrates expression [2.37].

5.2.8.– **(Chi² fitting test)** (see p. 58) Type the program:

```
# -*- coding: utf-8 -*-
"""
Created on Wed Jun  1 09:42:51 2016
```

```
****** chi2test
@author: maurice
"""
from numpy import zeros, std, sort, inf, nansum, diff
from numpy.random import rand, randn
from scipy.stats import norm, chi2
from matplotlib import pyplot as plt
nbgroups = 8; nbvalpergroup = 30; N = nbgroups*nbvalpergroup;
Lruns = 30000; Tchi2 = zeros(Lruns);
for irun in range(Lruns):
    sigma = 3*rand(); x = sigma*randn(N);
    hatsigma = std(x); xsort = sort(x);
    intervbounds = zeros(nbgroups+1); intervbounds[0] = -inf;
    for ig in range(nbgroups):
        aux_bound = (ig+1)*nbvalpergroup;
        intervbounds[ig+1] = xsort[aux_bound-1];
    cdfj = norm.cdf(intervbounds,0,hatsigma); npj = diff(cdfj)*
    float(N);
    X2 = nansum(((nbvalpergroup-npj)**2) / npj); Tchi2[irun] = X2;
bins = 50; plt.clf()
auxhist = plt.hist(Tchi2,bins=bins,normed='True',histtype=
'stepfilled');
plt.hold('on')
xvallin = auxhist[1][0:bins]+(auxhist[1][1]-auxhist[1][0])/2.0;
plt.plot(xvallin,chi2.pdf(xvallin,nbgroups-1),'.-m')
plt.hold('off'); plt.grid('on'); plt.show()
```

5.2.9.– (CRB expression using symbolic calculus) (see p. 61)

$$m = \begin{bmatrix} m_1 \\ m_2 \end{bmatrix} \text{ and } C = \begin{bmatrix} \sigma_1^2 & \rho\sigma_1\sigma_2 \\ \rho\sigma_1\sigma_2 & \sigma_2^2 \end{bmatrix}$$

Hence, $\partial_\rho m = 0$ and

$$\partial C_{\sigma_1} = \begin{bmatrix} 2\sigma_1 & \rho\sigma_2 \\ \rho\sigma_2 & 0 \end{bmatrix} \quad \partial C_{\sigma_2} = \begin{bmatrix} 0 & \rho\sigma_1 \\ \rho\sigma_1 & 2\sigma_2 \end{bmatrix} \quad \partial C_\rho = \begin{bmatrix} 0 & \sigma_1\sigma_2 \\ \sigma_1\sigma_2 & 0 \end{bmatrix}$$

Because the mean and the covariance depend on different parameters, the FIM is 2-block diagonal, one of size 2×2 associated with the mean and the other of size 3×3 associated with σ_1, σ_2 and ρ.

The following program performs the CRB. The coefficient of ρ has the very simple form $(1 - \rho^2)^2/N$. Type and run the program:

```
# -*- coding: utf-8 -*-
"""
Created on Sun Aug  7 16:16:55 2016
****** symbolicforCRBGaussian
@author: maurice
"""
from numpy import zeros
import sympy as sp
m1=sp.Symbol('m1'); m2=sp.Symbol('m2'); m=sp.Matrix([[m1],[m2]])
dm1 = sp.diff(m,m1); dm2 = sp.diff(m,m2)
s1 = sp.Symbol('s1'); s2 = sp.Symbol('s2'); r  = sp.Symbol('r')
C  = sp.Matrix([[s1**2, r*s1*s2],[r*s1*s2, s2**2]])
dCs1 = sp.diff(C,s1); dCs2 = sp.diff(C,s2); dCr = sp.diff(C,r)
iC = sp.Inverse(C)
F00m=(dm1.T*iC*dm1)[0];F01m=(dm1.T*iC*dm2)[0];F11m=(dm2.T*iC*dm2)[0]
Fmean = sp.Matrix([[F00m,F01m],[F01m,F11m]])
iCdCs1 = iC*dCs1; iCdCs2 = iC*dCs2; iCdCr  = iC*dCr
F00c=sp.trace((iCdCs1*iCdCs1)/2); F01c=sp.trace((iCdCs1*iCdCs2)/2)
F02c=sp.trace((iCdCs1*iCdCr)/2); F11c=sp.trace((iCdCs2*iCdCs2)/2)
F12c=sp.trace((iCdCs2*iCdCr)/2); F22c=sp.trace((iCdCr*iCdCr)/2)
Fcovar=sp.Matrix([[F00c,F01c,F02c],[F01c,F11c,F12c],[F02c,F12c,F22c]])
F=sp.Matrix(zeros([5,5])); F[0:2,0:2]=Fmean; F[2:5,2:5] = Fcovar
CRB = sp.Inverse(F)
print(sp.simplify(CRB[0,0])); print(sp.simplify(CRB[1,1]))
print(sp.simplify(CRB[2,2])); print(sp.simplify(CRB[3,3]))
print(sp.simplify(CRB[4,4]))
```

5.2.10.– (Decomposition of the design matrix) (see p. 68)

1) As Π_1 and Π_{X_c} are orthogonal projectors, $(\Pi_1 + \Pi_{X_c})$ is a projector. To show that $\Pi_Z = \Pi_1 + \Pi_{X_c}$, we must simply demonstrate that $(\Pi_1 + \Pi_{X_c})Z = Z$. Using the fact that $Z = \begin{bmatrix} \mathbb{1}_N & X \end{bmatrix}$ we obtain successively:

$$(\Pi_1 + \Pi_{X_c})Z = (\Pi_1 + \Pi_{X_c}) \begin{bmatrix} \mathbb{1}_N & X \end{bmatrix}$$

$$= \begin{bmatrix} \Pi_1 \mathbb{1}_N & \Pi_1 X \end{bmatrix} + \begin{bmatrix} \Pi_{X_c} \mathbb{1}_N & \Pi_{X_c} X \end{bmatrix}$$

$$= \begin{bmatrix} \mathbb{1}_N & \Pi_1 X \end{bmatrix} + \begin{bmatrix} 0 & X_c \end{bmatrix} = Z$$

where, following equations [2.63] and [2.64], we use the fact that $X_c = \Pi_{X_c} X_c = \Pi_{X_c} X + 0$.

2) As Π_1, Π_{X_c} and Π_Z are projectors and $\Pi_Z = \Pi_1 + \Pi_{X_c}$, we have, for any $v \in \mathbb{C}^N$:

$$v^H v \geq v^H \Pi_Z v = v^H \Pi_1 v + v^H \Pi_{X_c} v \geq v^H \Pi_1 v$$

In other words, $I_N \geq \Pi_Z \geq \Pi_1$.

3) We have $u_n^T \Pi_1 u_n = 1/N$, $u_n^T u_n = 1$ and $u_n^T \Pi_Z u_n = h_{n,n}$.

4) Type and run the program leverageeffect.py. Changing the test value, 0 or 1, we observe the effect of significant leverage.

```
# -*- coding: utf-8 -*-
"""
Created on Fri Jun  3 13:21:35 2016
****** leverageeffect
@author: maurice
"""
from numpy import mean, ones, zeros, dot
from numpy.random import randn
from numpy.linalg import inv
from matplotlib import pyplot as plt
N=100; P=2; X = randn(N,P);
if 0:
    X[N-1,:] = mean(X[range(N-1),:],axis=0);
else:
    X[N-1,:] = 100.0*mean(X[range(N-1),:],axis=0);
Z = zeros([N,P+1])
Z[:,0] = ones(N); Z[:,1:P+1]=X;
PiZ = dot(dot(Z,inv(dot(Z.transpose(),Z))),Z.transpose())
plt.clf(); plt.plot(X[:,0],X[:,1],'x'); plt.hold('on')
plt.plot(X[N-1,0],X[N-1,1],'or'); plt.hold('off')
plt.title('%4.2f<h_N=%4.2f =<1'%(1.0/N,PiZ[N-1,N-1]));plt.show()
```

5.2.11.– **(Atmospheric CO_2 concentration)** (see p. 68)

```
# -*- coding: utf-8 -*-
"""
Created on Mon Jun 20 17:18:19 2016
****** co2linmod
@author: maurice
"""
from numpy import zeros, arange, pi, cos, sin
from numpy import isnan, array, sqrt
import statsmodels.api as sm
```

```
from numpy import matrix as mat
from numpy.linalg import pinv
from matplotlib import pyplot as plt
dataCO2 = sm.datasets.co2.load(); y = dataCO2.data['co2']
date = dataCO2.data['date']; N = len(y)
# determine nan value indices
listindex=list([])
for ip in range(N):
    if not isnan(y[ip]):
        listindex.append(ip)
t=arange(N); tprime=t[listindex]; tmat = mat(tprime).transpose()
yprime = y[listindex]; Nmat = len(yprime); ymat = mat(yprime).
transpose()
q=3; p=2; f0 = 7.0/365.0; H = mat(zeros([Nmat,2*p+q]))
theta = 2*pi*f0*tmat*mat(arange(1,p+1))
H[:,0:p] = cos(theta); H[:,p:2*p] = sin(theta)
for iq in range(q):
    H[:,2*p+iq] = mat(tprime**iq).transpose()
pinvH = pinv(H); alpha = pinvH*ymat; ymatpred = H*alpha
e = ymat-ymatpred; stde = sqrt(e.transpose()*e/Nmat)
plt.clf()
plt.subplot(211); plt.plot(array(ymat)); plt.plot(array(ymatpred))
plt.subplot(212); plt.plot(array(e));
plt.title('std of the residue std = %4.2f'%(stde)); plt.show()
```

5.2.12.– **(Change-point detection of Nile flow)** (see p. 69) Type the program

```
# -*- coding: utf-8 -*-
"""
Created on Wed Jun 22 22:38:57 2016
****** modlinchangeNILE
@author: maurice
"""
from numpy import zeros, mean, arange, ones, array
from numpy import matrix as mat
from scipy.linalg import pinv
import statsmodels.api as sm
from matplotlib import pyplot as plt
datanile = sm.datasets.nile.load(); x = datanile.data['volume'];
year = datanile.data['year']
N=len(x); err = zeros(N-1); alpha0 = mat(zeros([2,N-1])); alpha1
= mat(zeros([2,N-1]))
for n in range(1,N):
    H0 = mat(zeros([n,2])); H0[:,0]=ones([n,1]);
```

```
    H0[:,1] = arange(n).reshape(n,1); pinvH0 = pinv(H0);
    x0 = mat(x[range(n)]); alpha0[:,n-1] = pinvH0*x0.transpose()
    err0 = x0*x0.transpose()-x0*H0*alpha0[:,n-1]
    H1 = mat(zeros([N-n,2])); H1[:,0]=ones([N-n,1]);
    H1[:,1] = arange(n,N).reshape(N-n,1); pinvH1 = pinv(H1)
    x1 = mat(x[range(n,N)]); alpha1[:,n-1] = pinvH1*x1.transpose()
    err1 = x1*x1.transpose() - x1*H1*alpha1[:,n-1]; err[n-1] =
    err0+err1
nopt = err.argmin()+1
alpha0opt = array(alpha0[:,nopt-1]); alpha1opt = array(alpha1
[:,nopt-1])
mu0opt = mean(x[0:nopt]); mu1opt = mean(x[nopt:N])
plt.clf(); plt.subplot(211); plt.plot(year,x)
plt.hold('on'); plt.plot((year[0], year[nopt]),\
      (alpha0opt[0], alpha0opt[0]+alpha0opt[1]*nopt),'--')
plt.plot((year[nopt], year[N-1]),\
      (alpha1opt[0], alpha1opt[0]+alpha1opt[1]*nopt),'--')
plt.hold('off')
plt.xlim([year[0],year[N-1]])
plt.xticks(fontsize = 8); plt.yticks(fontsize = 8)
plt.subplot(212); plt.plot(year[range(1,N)], err,'.-')
plt.hold('on'); plt.xlim([year[0],year[N-1]])
plt.plot(year[nopt], err[nopt-1],'or'); plt.hold('off')
plt.yticks([]); plt.xticks(fontsize = 8); plt.yticks(fontsize = 8)
plt.title('change in %i, mu0 = %4.2f, mu1 = %4.2f'%(year[nopt],
alpha0opt[0],\
    alpha1opt[0]), fontsize = 8)
plt.show()
```

5.2.13.– **(Confidence intervals on linear model features)** (see p. 79) Type and run the following program:

```
# -*- coding: utf-8 -*-
"""
Created on Wed Jul 27 11:28:56 2016
****** regagediabetes
@author: maurice
"""
from numpy import zeros, ones, sqrt, diag, dot
from numpy import min, max, linspace
from numpy.linalg import pinv
from scipy.stats import t
from sklearn import datasets as ds
import matplotlib.pyplot as plt
```

```
diabetes = ds.load_diabetes();
expl = diabetes['data']; target = diabetes['target'][0:30]
N = len(target); X = expl[0:N,0].reshape(N); y = target.reshape(N)
Xsort = X[X.argsort()]; ysort = y[X.argsort()]
Z = zeros([N,2]); Z[:,0] = ones(N); Z[:,1] = Xsort
iZTZ=pinv(dot(Z.transpose(),Z)); pinvZ = dot(iZTZ,Z.transpose());
hatbeta=dot(pinvZ, ysort); haty = dot(Z,hatbeta); e = ysort-haty;
hatsigma2=(e.transpose()*e)/(N-2); hatsigma2 = dot(e.transpose(),
e)/(N-2)
alphaIC = 0.05; STDhatbeta=sqrt(hatsigma2*diag(iZTZ));
ICZbeta = zeros(N); hi = zeros(N);
for ii in range(N):
    hi[ii] = dot(Z[ii,:],pinvZ[:,ii]);
    ICZbeta[ii] = t.isf(alphaIC/2.0,N-2)*sqrt(hatsigma2*hi[ii]);
plt.clf(); plt.plot(Xsort,ysort,'go',label='true data')
plt.hold('on'); plt.plot(Xsort,haty-ICZbeta,'r.-', \
    label='prediction on learning observations');
plt.plot(Xsort,haty,'r.--'); plt.plot(Xsort,haty+ICZbeta,'r.-');
nbnewvalues = 12; listZ2o = linspace(min(X),max(X),nbnewvalues)
hatyo = zeros(nbnewvalues); IChatyo = zeros(nbnewvalues)
for iZ2o in range(nbnewvalues):
    Zo = [1.0,listZ2o[iZ2o]]; hatyo[iZ2o] = dot(Zo,hatbeta);
    h0 = dot(Zo,dot(iZTZ,Zo));
    IChatyo[iZ2o] = t.isf(alphaIC/2,N-2)*sqrt(hatsigma2*(1+h0));
plt.plot(listZ2o,hatyo-IChatyo,'b.-',label='prediction on new
observations');
plt.plot(listZ2o,hatyo,'b.--');plt.plot(listZ2o,hatyo+IChatyo,'b.-')
plt.hold('off');plt.legend(loc='best',fontsize=12);
plt.xlabel('observation'); plt.ylabel('target'); plt.show()
```

5.2.14.– (**Hypothesis test on** $H_0 = \{\beta_{1:p} = 0\}$) (see p. 79) Type and run the following program:

```
# -*- coding: utf-8 -*-
"""
Created on Wed Jun 15 06:53:26 2016
****** regboston
@author: maurice
"""
from sklearn import datasets
from numpy import sqrt, zeros, ones, mean, array
from numpy import matrix as mat
from numpy.linalg import pinv
from scipy.stats import f
```

```
bostondata = datasets.load_boston()
print(bostondata.DESCR[:1200])
Y = bostondata.target; X = bostondata.data; C = bostondata.feature
_names
ourselection = array(['RM','LSTAT','CRIM','ZN','CHAS','DIS'])
N=len(Y); P=len(ourselection); selectedC=zeros(len(ourselection))
Xselect = zeros([N,P]); cp = 0;
for ip in range(len(C)):
    if (ourselection==C[ip]).any():
        Xselect[:,cp] = X[:,ip]; cp = cp+1;
Z = mat(zeros([N,P+1])); Z[:,0] = ones([N,1]); Z[:,range(1,P+1)] =
Xselect
y = mat(Y).reshape(N,1); iZTZ = pinv(Z.transpose()*Z); pinvZ =
iZTZ * Z.transpose();
hatbeta = pinvZ * y; haty = Z*hatbeta; e = y-haty;
hatsigma2 = (e.transpose()*e)/(N-P-1); hatsigma = sqrt(hatsigma2);
#===== test on beta_1=...=beta_P=0
RSS = (e.transpose()*e); ycentered = (y-mean(y))
TSS = ycentered.transpose()*ycentered
ESS = TSS-RSS; F = (N-P-1)*ESS/(P*RSS)
pvalueF = 1.0-f.cdf(F,P,N-P-1)
print('******* our selection %s'%ourselection)
print('******* pvalueF = %4.2e'%pvalueF)
```

5.2.15.– **(Model selection based on Z-score)** (see p. 79) Type and run the following program:

```
# -*- coding: utf-8 -*-
"""
Created on Sun Jul 31 22:23:34 2016
****** diabetesZscore
@author: maurice
"""
from numpy import zeros, ones, dot, diag, loadtxt, sqrt, mean, std
from numpy import argwhere
from numpy.linalg import pinv
from scipy.stats import t
from sklearn import datasets as ds
if 0:
    diabetes = ds.load_diabetes();
    Xnonstandard = diabetes['data']; y = diabetes['target']
    N,p = Xnonstandard.shape; r = p+1;
else:
    P = loadtxt('prost.txt'); # page 63 Tibshirani
```

```
    N,q = P.shape; p = q-1; r = p+1;
    Xnonstandard = P[:,0:p]; y = P[:,p]
standardizatioflag = False
if standardizatioflag:
    Xc = Xnonstandard-dot(ones([N,1]),mean(Xnonstandard,0).reshape
    (1,p))
    X = Xc / dot(ones([N,1]),std(Xc,0).reshape(1,p))
else:
    X = Xnonstandard
Z = zeros([N,r]); Z[:,0] = ones(N); Z[:,1:r] = X;
diZTZ = diag(pinv(dot(Z.T,Z))); pinvZ = pinv(Z); hatbeta = dot
(pinvZ, y);
haty = dot(Z,hatbeta); e = haty-y; hatsigma2 = sum(e**2)/(N-r);
Zscore = hatbeta/sqrt(diZTZ)/sqrt(hatsigma2)
Zscoreintercept = Zscore[0]; Zscoreexplanatory = Zscore[1:]
alphaIC = 0.05; threshold = t.isf(alphaIC/2,N-r)
print(argwhere(abs(Zscoreexplanatory)>threshold))
```

5.2.16.– (Model selection based on *adjusted* R^2**, AIC and BIC)** (see p. 81)

```
# -*- coding: utf-8 -*-
"""
Created on Wed Jul 27 11:28:56 2016
****** diabetesvalidationmodel
@author: maurice
"""
from numpy import zeros, ones, dot, setdiff1d, mean, log
from numpy.linalg import pinv
from sklearn import datasets as ds
import matplotlib.pyplot as plt
def validationfeatures(X,y):
    N,p = X.shape; r = p+1;
    H =zeros([N,r]); H[:,0] = ones(N); H[:,1:r] = X;
    invH = pinv(H); hatbeta = dot(invH,y);
    haty = dot(H,hatbeta);
    ESS = sum((haty-mean(y)) **2); TSS = sum((y-mean(y)) **2);
    RSS = TSS-ESS; adjR2 = 1.0-(RSS/(N-r)/(TSS/(N-1)))
    AIC = N*log(RSS/N)+2.0*r; BIC = N*log(RSS/N)+r*log(N);
    return adjR2, BIC, AIC
def residue_norm(X,y):
    N,p = X.shape; r = p+1;
    Z = zeros([N,r]); Z[:,0] = ones(N); Z[:,1:r] = X;
    pinvZ = pinv(Z); hatbeta = dot(pinvZ, y);
    haty = dot(Z,hatbeta); T = sum((haty-y) **2)
```

```
    return T
#========== main program ====
diabetes = ds.load_diabetes(); explanatoryVars = diabetes['data']
target = diabetes['target']; N,p = explanatoryVars.shape;
X = explanatoryVars.reshape(N,p); y = target.reshape(N,1)
aR2 = zeros(p+1); BIC = zeros(p+1); AIC = zeros(p+1);
aR2[p], BIC[p], AIC[p] = validationfeatures(X,y)
T = zeros(p+1); T[p] = residue_norm(X,y)
suppressedVarindex = zeros(p);
model_k = X; colmodel_kindex = range(p)
for k in range(p,0,-1):
    T_k = zeros(k); rangek = range(k);
    for j in rangek:
        testmodel_jk = model_k[:,setdiff1d(rangek,j)]
        T_k[j] = residue_norm(testmodel_jk,y)
    jo = T_k.argmin()
    T[k-1] = T_k.min()
    suppressedVarindex[k-1] = colmodel_kindex[jo]
    colmodel_kindex=setdiff1d(colmodel_kindex,colmodel_kindex[jo])
    model_k = model_k[:,setdiff1d(rangek,jo)]
    aR2[k-1], BIC[k-1], AIC[k-1] = validationfeatures(model_k,y)
plt.clf();
plt.subplot(3,1,1); plt.plot(aR2,'.-'); plt.show()
plt.subplot(3,1,2); plt.plot(BIC,'.-'); plt.show()
plt.subplot(3,1,3); plt.plot(AIC,'.-'); plt.show()
selectedcolumnsaR2 = setdiff1d(range(p),suppressedVarindex[aR2.
argmax():p])
print('***** The optimal model consists of the %i explanatory
features : %s' \
 %(len(selectedcolumnsaR2),selectedcolumnsaR2))
selectedcolumnsBIC = setdiff1d(range(p),suppressedVarindex[BIC.
argmin():p])
print('***** The optimal model consists of the %i explanatory
features : %s' \
 %(len(selectedcolumnsBIC),selectedcolumnsBIC))
selectedcolumnsAIC = setdiff1d(range(p),suppressedVarindex[AIC.
argmin():p])
print('***** The optimal model consists of the %i explanatory
features : %s' \
 %(len(selectedcolumnsAIC),selectedcolumnsAIC))
```

5.2.17.– (**Moment estimator: central limit theorem**) (see p. 82) Let $\widehat{m}_1 = N^{-1}\sum_{n=0}^{N-1} X_n$ and $\widehat{m}_2 = N^{-1}\sum_{n=0}^{N-1} X_n^2$. Expression [2.100] may be rewritten $\begin{bmatrix} \widehat{k} & \widehat{\lambda} \end{bmatrix}^T = g(\widehat{m}_1, \widehat{m}_2)$, where:

$$g: \begin{bmatrix} m_1 \\ m_2 \end{bmatrix} \longrightarrow \begin{bmatrix} k \\ \lambda \end{bmatrix} = \begin{bmatrix} \dfrac{m_1^2}{m_2 - m_1^2} \\ \dfrac{m_2 - m_1^2}{m_1} \end{bmatrix} \quad [5.7]$$

$$\Leftrightarrow g^{-1}: \begin{bmatrix} k \\ \lambda \end{bmatrix} \longrightarrow \begin{bmatrix} m_1 \\ m_2 \end{bmatrix} = \begin{bmatrix} k\lambda \\ k(1+k)\lambda^2 \end{bmatrix}$$

The central limit theorem 1.9 states that when N tends toward infinity:

$$\sqrt{N} \begin{bmatrix} \widehat{m}_1 - k\lambda \\ \widehat{m}_2 - k(1+k)k^2 \end{bmatrix} \to \mathcal{N}(0, C(\alpha, \lambda))$$

where:

$$C(k, \lambda) = 2 \begin{bmatrix} \mathrm{cov}(X_n, X_n) & \mathrm{cov}(X_n, X_n^2) \\ \mathrm{cov}(X_n, X_n^2) & \mathrm{cov}(X_n^2, X_n^2) \end{bmatrix}$$

Readers may wish to determine the expression of $C(k, \lambda)$ as a function of k and λ. The continuity theorem 1.10, applied to function g, states that when N tends toward infinity:

$$\sqrt{N} \begin{bmatrix} \widehat{k} - k \\ \widehat{\lambda} - \lambda \end{bmatrix} \to \mathcal{N}(0, \Gamma(\alpha, \lambda))$$

where $\Gamma(k, \lambda) = JCJ^T$ with:

$$J = \begin{bmatrix} \dfrac{\partial g_1}{m_1} & \dfrac{\partial g_1}{m_2} \\ \dfrac{\partial g_2}{m_1} & \dfrac{\partial g_2}{m_2} \end{bmatrix}$$

This may be calculated in terms of (k, λ) using [5.7]. Clearly, the performance depends on the parameter values. In the context of a practical problem, if we wish to calculate a confidence interval, the values of the parameter (k, λ) may be replaced by an estimate.

5.2.18.– (Moment estimators of mixture proportion) (see p. 83)

1) The statistical model has a density of:

$$p(x_0, \ldots, x_{N-1}) = \prod_{n=0}^{N-1} \left(\frac{\alpha}{\sigma_0 \sqrt{2\pi}} e^{-(x_n - m_0)^2 / 2\sigma_0^2} + \frac{1-\alpha}{\sigma_1 \sqrt{2\pi}} e^{-(x_n - m_1)^2 / 2\sigma_1^2} \right)$$

where the parameter $\alpha \in (0, 1)$. We deduce that $\mathbb{E}\{X_n\} = \alpha m_0 + (1-\alpha) m_1$ and $\mathbb{E}\{X_n^2\} = \alpha(m_0^2 + \sigma_0^2) + (1-\alpha)(m_1^2 + \sigma_1^2)$.

2) We have $S(\alpha) = \mathbb{E}\{S(X)\} = \alpha m_0 + (1-\alpha) m_1 \Rightarrow \alpha = (S(\alpha) - m_1)/(m_0 - m_1)$. Hence, $\widehat{\alpha}_1 = (S(X) - m_1)/(m_0 - m_1)$.

3) $S(\alpha) = \mathbb{E}\{S(X)\}$. Therefore:

$$S(\alpha) = \begin{bmatrix} \alpha m_0 + (1-\alpha) m_1 \\ \alpha(m_0^2 + \sigma_0^2) + (1-\alpha)(m_1^2 + \sigma_1^2) \end{bmatrix}$$

Let $\widehat{S}(X) = \begin{bmatrix} \widehat{S}_0(X) & \widehat{S}_1(X) \end{bmatrix}$ and $J(\alpha) = \|S(\alpha) - \widehat{S}(X)\|^1$. From this, we deduce the value $\widehat{\alpha}_2$ that minimizes $J(\alpha)$, canceling the derivative of $J(\alpha)$.

4) Using the fact that for a Gaussian, centered, random variable U of variance σ^2, we have $\mathbb{E}\{U^3\} = 0$ and $\mathbb{E}\{U^4\} = 3\sigma^4$, we obtain:

$$C(\alpha) = \alpha \begin{bmatrix} \sigma_0^2 & 0 \\ 0 & 3\sigma_0^4 \end{bmatrix} + (1-\alpha) \begin{bmatrix} \sigma_1^2 & 0 \\ 0 & 3\sigma_1^4 \end{bmatrix}$$

We obtain the estimator $\widehat{\alpha}_3$ by minimizing the function $K(\alpha) = (\widehat{S}(X) - S(\alpha))^T C^{-1}(\alpha)(\widehat{S}(X) - S(\alpha))$ in relation to α.

5) Type and run the following program:

```
# -*- coding: utf-8 -*-
"""
Created on Fri Jun 3 06:32:09 2016
****** MMmixture
@author: maurice
"""
from numpy import linspace, zeros, dot, sum, mean, std, array
from numpy.linalg import pinv
from numpy.random import randn, rand
from matplotlib import pyplot as plt #======
#======
```

```
def fgmm(S,m0,m1,s0,s1,La):
    """
    Estimation of the mixture proportion alpha,
    using the two first 2 moments
    and by an exhaustive research on a grid of values of alpha.
    Rk: m0,m1,s0,s1 are assumed to be known
    synopsis:
    fgmm(S,m0,m1,s0,s1,La) inputs:
    S = data
    m0, m1 = means of the 2 components
    s0, s1 = standard deviation of the 2 components
    La = number of values on the alpha grid
    output:
        alphaopt = optimal proportion
    """
    F0=array([m0,m0**2+s0**2]);W0=array([[s0**2,0],[0, 3*s0**4]]);
    F1=array([m1,m1**2+s1**2]);W2=array([[s1**2,0],[0, 3*s1**4]]);
    alphalist=linspace(0,1,La);
    valt = zeros(La);
    for ia in range(La):
        alpha = alphalist[ia];
        W = alpha*W0+(1.0-alpha)*W2;
        M = alpha*F0+(1.0-alpha)*F1;
        PiW = pinv(W)
        valt[ia] = dot(dot((S-M).transpose(),PiW),(S-M));
    iamin = valt.argmin(); alphaopt=alphalist[iamin];
    return alphaopt
#============== main program ===============
alpha=0.5; m0=10; m1=10.1; s0=0.2; s1=0.5; N=100;
Lruns=1000; hatalpha=zeros([Lruns,3]);
La = 100;
for ir in range(Lruns):
    U=rand(N)<alpha;
    X=(m0+s0*randn(N))*U + (m1+s1*randn(N))*(1-U);
    #===== we use only the mean statistic
    S0=mean(X); F00=m0; F01=m1; hatalpha[ir,0]=(S0-F01)/(F00-F01);
    #===== we use the two first statistics
    S1=sum(X**2)/N; S=array([S0,S1]); F0=array([F00,m0**2+s0**2]);
    F1=array([F01,m1**2+s1**2]);
    H = F0-F1; pH = 1.0 / (H[0]**2+H[1]**2)
    hatalpha[ir,1] = ((S[0]-F1[0])*H[0]+ (S[1]-F1[1])*H[1])*pH;
    hatalpha[ir,2] = fgmm(S,m0,m1,s0,s1,La);
print(std(hatalpha-alpha,axis=0))
```

```
plt.clf(); plt.hold('off'); plt.boxplot(hatalpha); plt.show()
```

In this case, the minimization applies to the scalar variable $\alpha \in (0, 1)$. The minimization operation may therefore be carried out using a value grid. Modifying the parameters m_i, σ_i, we see that the reduction of the mean squares error obtained with $\widehat{\alpha}_2$ and $\widehat{\alpha}_3$ in relation to $\widehat{\alpha}_1$ depends to a significant extent on the choice of these values.

5.2.19.– **(MLE for the linear model)** (see p. 88) The model consists of N Gaussian random variables with respective means $Z_n \theta$ and same variance σ^2. Then, the log-likelihood writes:

$$\ell = -\frac{N}{2} \log(2\pi) - \frac{N}{2} \log \sigma^2 - \frac{1}{2\sigma^2} \sum_{n=0}^{N-1} (y_n - Z_n \theta)^2$$

$$= -\frac{N}{2} \log(2\pi) - \frac{N}{2} \log \sigma^2 - \|y - Z\theta\|^2$$

The cancellation of the first derivative with respect to θ gives $\widehat{\theta}_{\text{MLE}} = (Z^T Z)^{-1} Z^T y$ and leads to the maximum:

$$\widetilde{\ell} = -\frac{N}{2} \log(2\pi) - \frac{N}{2} \log(\sigma^2) - \frac{\text{RSS}}{2\sigma^2}$$

where RSS is given by [2.88]. Then, the cancellation of the first derivative with respect to σ^2 gives:

$$\partial_{\sigma^2} \widetilde{\ell} = -\frac{N}{2} \sigma^2 + \frac{1}{2} \sigma^4 \text{RSS} = 0 \Rightarrow \widehat{\sigma}^2_{\text{MLE}} = \text{RSS}/N$$

Hence, the MLE of θ is equal to the least square estimator [2.71], but $\widehat{\sigma}^2_{\text{MLE}}$ is different from the unbiased least square estimator [2.76].

5.2.20.– **(Iris classification)** (see p. 88) From section 2.6.5.1, we derive the expressions of the respective estimates of mean and covariance of the three models. We denote $\ell(x; \theta_i)$ the log-likelihood associated with class i. It follows that the maximum likelihood class associated with the observation x_o is given by:

$$\widehat{k} = \arg\max_i \left(-\log \det \{C_i\} - \text{trace} \{C_i^{-1} R_i\} \right)$$

where $R_i = N^{-1} \sum_{n=0}^{N-1} (x_o - m_i)(x_o - m_i)^T$ and m_i and C_i have been obtained from the training dataset with an LDA pre-processing with $k = 2$.

Type and run the following program where you can change the values of k, the respective sizes of the training and testing datasets and replace the log-likelihood by the Mahalanobis distance [2.119].

```
# -*- coding: utf-8 -*-
"""
Created on Mon Jul  4 16:05:14 2016
****** irisclassification
@author: maurice
"""
from numpy import zeros, log, mean, size, unique
from scipy.linalg import det, inv
from sklearn import datasets
from pcaldatoolbox import LDA
iris = datasets.load_iris()
X = iris.data; y = iris.target; t_names = iris.target_names
N = size(X,0); d = size(X,1); g = len(unique(y));
Ntrain = 35; Ntest = 50-Ntrain;
Xtrain = zeros([g*Ntrain,d]); ytrain = zeros([g*Ntrain])
Xtest  = zeros([g*Ntest,d]);  ytest  = zeros([g*Ntest])
k = 2; cpError = zeros(g); ell = zeros(g); cp = 0;
detC = zeros(g); ihatCtrain = zeros([k,k,g]); hatmtrain = zeros([k,g])
for ig in range(g):
    Xi = X[y==ig]; yi = y[y==ig]
    Xtrain[Ntrain*ig:Ntrain*(ig+1),:] = Xi[0:Ntrain,:];
    Xtest[Ntest*ig:Ntest*(ig+1),:] = Xi[Ntrain:N,:];
    ytrain[Ntrain*ig:Ntrain*(ig+1)] = yi[0:Ntrain];
    ytest[Ntest*ig:Ntest*(ig+1)] = yi[Ntrain:N];
XLDAtrain, VLDAtrain = LDA(Xtrain,ytrain,k);
for ig in range(g):
    XLDAtrain_ig = XLDAtrain[ytrain==ig,:]
    hatmtrain[:,ig] = mean(XLDAtrain_ig,0)
    Xigc = XLDAtrain_ig - hatmtrain[:,ig]
    hatC = Xigc.T*Xigc/Ntrain
    detC[ig] = det(hatC)
    ihatCtrain[:,:,ig] = inv(hatC)
for itest in range(g*Ntest):
    for ig in range(g):
        Xit = Xtest[itest,:]*VLDAtrain-hatmtrain[:,ig]
        ell[ig] = -log(detC[ig])-(Xit*ihatCtrain[:,:,ig])*Xit.T
    if not(ell.argmax()==ytest[itest]):
        cp = cp + 1
print('***** Prediction error number = %i'%cp)
```

5.2.21.– (**Asymptotic distribution of a correlation MLE**) (see p. 88)

1) $\rho = C_{0,1}/\sqrt{C_{0,0}C_{1,1}}$ appears as a function $f(C)$. Therefore, a MLE of ρ is given by $f(\widehat{C})$ that writes:

$$\widehat{\rho} = \frac{\widehat{C}_{0,1}}{\sqrt{\widehat{C}_{0,0}\widehat{C}_{1,1}}}$$

2) Type and run the following program:

```
# -*- coding: utf-8 -*-
"""
Created on Sun Aug  7 18:32:07 2016
****** verifyCRBrho
@author: maurice
"""
from numpy.random import randn
from numpy import array, zeros, mean, std, matrix as mat, sqrt
from scipy.linalg import sqrtm
N=100; d=2; Lruns=500; m0 = 2; m1 = 4; s0 = 2; s1 = 5; rho = 0.7;
C = array([[s0**2, rho*s0*s1],[rho*s0*s1, s1**2]])
m = array([m0,m1]); sqrtC = mat(sqrtm(C))
hats0 = zeros(Lruns); hats1 = zeros(Lruns); hatrho = zeros(Lruns)
for ell in range(Lruns):
    W = mat(randn(N,2)); X = W*sqrtC+m; meanX = mat(mean(X,0));
    hatC = (X-meanX).T*(X-meanX)/N;
    hats0[ell] = sqrt(hatC[0,0]); hats1[ell] = sqrt(hatC[1,1])
    hatrho[ell] = hatC[0,1]/(hats0[ell]*hats1[ell])
print('** s1: \tempirical std = %4.4e\n\ttheoretical std = %4.4e'\
        %(std(hats0),s0/sqrt(2*N)))
print('** s2: \tempirical std = %4.4e\n\ttheoretical std = %4.4e'\
        %(std(hats1),s1/sqrt(2*N)))
print('** r:\tempirical std = %4.4e\n\ttheoretical std = %4.4e'\
        %(std(hatrho),(1-rho**2)/sqrt(N)))
```

5.2.22.– (**MM versus MLE of the correlation**) (see p. 89)

1) The model writes { i.i.d. $\mathcal{N}(N; 0, C)$} and the parameter is ρ in $(-1, +1)$. This model is different from the model described by [2.110] for $d = 2$. Indeed here, the two means and the two variances are assumed to be known. The log-likelihood writes:

$$\ell = -N\log(2\pi) - \frac{N}{2}\log(1-\rho^2) - \frac{1}{2(1-\rho^2)}\sum_{n=0}^{N-1}(X_{n,0}^2 + X_{n,1}^2 - 2\rho X_{n,0}X_{n,1})$$

2) From $\rho = C_{0,1}/\sqrt{C_{0,0}C_{1,1}}$, we derive an MM estimator $\widehat{\rho}_{\text{MM}} = S_{01}/\sqrt{S_{00}S_{11}}$, where $S_{00} = N^{-1}\sum_{n=0}^{N-1} X_{n,0}^2$, $S_{11} = N^{-1}\sum_{n=0}^{N-1} X_{n,1}^2$ and $S_{01} = N^{-1}\sum_{n=0}^{N-1} X_{n,0}X_{n,1}$.

3) Canceling the derivative of the log-likelihood with respect to ρ yields the cubic equation[1]:

$$\widehat{\rho}_{\text{MLE}}^3 - S_{01}\widehat{\rho}_{\text{MLE}}^2 + (S_{00} + S_{11} - 1)\widehat{\rho}_{\text{MLE}} - S_{01} = 0 \qquad [5.8]$$

Therefore, usually $\widehat{\rho}_{\text{MLE}} \neq \widehat{\rho}_{\text{MM}}$.

4) Type and run the following program:

```
# -*- coding: utf-8 -*-
"""
Created on Sun Aug  7 18:32:07 2016
****** MMversusMLEcorrelation
@author: maurice
"""
from numpy.random import randn
from numpy import array, zeros, nan, roots, nanstd, matrix as mat
from numpy import sqrt, isreal, isnan
from scipy.linalg import sqrtm
d=2; rho = 0.8; N = 200; Lruns = 500;
C = array([[1, rho],[rho, 1]]); sqrtC = mat(sqrtm(C))
hatrhoMM = zeros(Lruns); hatrhoMLE = zeros(Lruns)
for ell in range(Lruns):
    W = mat(randn(N,2)); X = W*sqrtC; hatC = X.T*X;
    S01 = hatC[0,1]/N; S00 = hatC[0,0]/N; S11 = hatC[1,1]/N;
    hatrhoMM[ell] = S01/sqrt(S00*S11)
    coeffpoly3nd = array([1.0, -S01, (S00+S11-1.0), -S01])
    rootsMLE = roots(coeffpoly3nd)
    realroots = rootsMLE[isreal(rootsMLE)].real
    if realroots.ndim==1:
        hatrhoMLE[ell] = realroots;
    else:
        hatrhoMLE[ell] = nan;
```

[1] If the discriminant $\Delta = 18abcd - 4b^3 d + b^2 c^2 - 4ac^3 - 27a^2 d^2$ of the cubic equation $ax^3 + bx^2 + cx + d = 0$ is strictly less than 0, then the cubic equation has one real root and two non-real conjugate roots.

```
print('\tstd of MM estimator=%4.3f\n\tstd of MLE estimator=%4.3f'\
        %(nanstd(hatrhoMM),nanstd(hatrhoMLE)))
print('\tnumber of real multiple roots=%i'%sum(isnan(hatrhoMLE)))
```

5.2.23.– (Correlation GLRT) (see p. 89)

1) We denote $X_k = \begin{bmatrix} X_{k,0} & X_{k,1} \end{bmatrix}^T$. Then, we can apply the results of section 2.6.5.1 with $d = 2$. The parameter of interest is $\theta = (m_0, m_1, \sigma_0, \sigma_1, \rho) \in \Theta = \mathbb{R} \times \mathbb{R} \times \mathbb{R}^+ \times \mathbb{R}^+ \times (-1, 1)$ and an **MLE** estimator of ρ writes $\widehat{\rho} = \widehat{C}_{0,1}/\sqrt{\widehat{C}_{0,0}\widehat{C}_{1,1}}$, where

$$\widehat{C} = \frac{1}{n}\sum_{k=0}^{n-1}(X_n - \widehat{m})(X_n - \widehat{m})^T \text{ with } \widehat{m} = \frac{1}{n}\sum_{k=0}^{n-1} X_n$$

It is easy to show that the log-likelihood maximum, given by [2.113], is an increasing function of the statistic $|\widehat{\rho}|^2$. Therefore, the GLRT statistic of $H_0 = \{\theta : |\rho| \leq \rho_0\}$ writes $|\widehat{\rho}|/\rho_0$, leading to the test:

$$|\widehat{\rho}| \underset{H_0}{\overset{H_1}{\gtrless}} \eta\rho_0, \text{ where } |\rho| = \frac{|\widehat{C}_{01}|}{\sqrt{\widehat{C}_{00}\widehat{C}_{11}}}$$

2) Since that the Fisher transform is monotonic, the test is equivalent to a test based on the statistic $\widehat{f} = 0.5\log((1+\widehat{\rho})/(1-\widehat{\rho}))$. The threshold can be determined using that, under H_0, $\widehat{f}$ is approximately distributed as a Gaussian with mean f_0 and variance $(n-3)^{-1}$.

3) The following program gives a p-value of 0.0176. We can therefore reject H_0. This indicates that the correlation is likely to have a modulus greater than 0.7.

```
# -*- coding: utf-8 -*-
"""
Created on Wed Jun  8 08:53:20 2016
****** testcorrelationWH
@author: maurice
"""
from numpy import mean, sqrt, log, sum
from scipy.stats import norm
Hcm = [162,167,167,159,172,172,168]; N = len(Hcm);
Wkg = [48.3,50.3,50.8,47.5,51.2,51.7,50.1];
Xc = Hcm-mean(Hcm);Yc = Wkg-mean(Wkg);
```

```
hatcorr = abs(sum(Xc*Yc)) / sqrt(sum(Xc*Xc)*sum(Yc*Yc));
hatf = 0.5*log((1.0+hatcorr)/(1.0-hatcorr));
rho0 = 0.7; f0 = 0.5*log((1+rho0)/(1-rho0));
pvalue = 2*(1-norm.cdf(hatf,f0,1/sqrt(N-3)));
print('***** p-value = %4.4f'%pvalue)
```

As an additional note, the following program shows that the distribution of the Fisher transformation of $\hat{\rho}$ approximately follows a Gaussian distribution of mean f_0 and variance $1/(N-3)$. We also compare to the asymptotic distribution derived from the CRB, see exercie 2.21.

```
# -*- coding: utf-8 -*-
"""
Created on Wed Jun  1 12:01:38 2016
****** correlationdistribution
@author: maurice
"""
from numpy import mean, log, sqrt, zeros, matrix as mat
from numpy.random import randn
from scipy.stats import norm
from scipy.linalg import sqrtm
from matplotlib import pyplot as plt
m = mat([1,2]); s0 = 5; s1 = 2; rho = -0.9; N = 20;
Lruns=10000; hatrho=zeros(Lruns); hatf=zeros(Lruns);
C = mat([[s0*s0, rho*s0*s1],[rho*s0*s1, s1*s1]]);
racC = sqrtm(C);
for ir in range(Lruns):
    Z = mat(randn(N,2)); X = Z*racC+m;
    Xc = X-mean(X,0); hatC = Xc.T*Xc/N
    hatrho[ir] = hatC[1,0] / sqrt(hatC[0,0]*hatC[1,1])
    hatf[ir] = 0.5*log((1.0+hatrho[ir])/(1.0-hatrho[ir]));
f0 = 0.5*log((1+rho)/(1-rho));
plt.clf(); bins = 30;
plt.subplot(121)
aux=plt.hist(hatrho,bins,normed='True');
plt.hold('on')
xtheo = aux[1][0:bins]+(aux[1][1]-aux[1][0])/2.0
plt.plot(xtheo,norm.pdf(xtheo,rho,(1-rho*rho)/sqrt(N)),'.-r')
plt.hold('off'); plt.show(); plt.title('asymptotic distribution')
plt.subplot(122)
aux=plt.hist(hatf,bins,normed='True');
```

```
plt.hold('on')
xtheo = aux[1][0:bins]+(aux[1][1]-aux[1][0])/2.0
plt.plot(xtheo,norm.pdf(xtheo,f0,1/sqrt(N-3)),'.-r')
plt.hold('off'); plt.show(); plt.title('fisher transform')
```

5.2.24.– (MLE with $\Gamma(k, \lambda)$ distribution) (see p. 90)

1) The $\Gamma(1, \lambda)$ distribution is an exponential distribution of parameter λ. The log-likelihood is expressed $\mathcal{L}(\lambda) = -N \log(\lambda) - \frac{1}{\lambda} \sum_{n=0}^{N-1} X_n$. Canceling the derivative in relation to λ, we obtain:

$$\widehat{\lambda} = \frac{1}{N} \sum_{n=0}^{N-1} X_n \quad [5.9]$$

The Fisher information is written as:

$$F = -\mathbb{E}\left\{ N/\lambda^2 - 2S/\lambda^3 \right\} = N/\lambda^2$$

Hence, following [2.108]:

$$\sqrt{N}(\widehat{\lambda} - \lambda) \xrightarrow{d} \mathcal{N}(0, \lambda^2)$$

Type and run the following program:

```
# -*- coding: utf-8 -*-
"""
Created on Sun Jun  5 07:24:22 2016
****** MLEexponential
@author: maurice
"""
from numpy import zeros, log, sqrt, std, mean
from numpy.random import rand
lamb=2; N=100; Lruns=300;
hatlamb=zeros(Lruns);
for ir in range(Lruns):
    Y = -lamb*log(rand(N));
    hatlamb[ir] = mean(Y);
print('**** std = %4.2f, theo-value = %4.2f'\
    %(std(hatlamb-lamb), lamb/sqrt(N)))
```

This program uses the property of inversion of the cumulative distribution function $F(x) = 1 - e^{-x/\lambda}$, and thus $x = -\lambda \log(1 - F)$, as presented in section 4.3.1. Finally, if F is a uniform r.v., then $(1 - F)$ is also a uniform r.v. This explains the generation of an exponential law generation using `Y=-lamb*log(rand(N));`.

2) The log-likelihood is written as:

$$\mathcal{L}(\theta) = -Nk\log\lambda - N\log\Gamma(k) + \sum_{n=0}^{N-1}(k-1)\log X_n - \sum_{n=0}^{N-1}\frac{X_n}{\lambda}$$

Canceling the Jacobian with respect to (k,λ), we obtain:

$$\begin{cases} -\dfrac{Nk}{\lambda} + \dfrac{1}{\lambda^2}\sum_{n=0}^{N-1} X_n = 0 \\ -N\log(\lambda) - N\dfrac{\Gamma'(k)}{\Gamma(k)} + \sum_{n=0}^{N-1}\log X_n = 0 \end{cases}$$

From the first equation, let $\widehat{\lambda} = \frac{1}{Nk}\sum_{n=0}^{N-1} X_n = S/k$. Applying this to the second equation gives us:

$$k - \frac{\Gamma'(k)}{\Gamma(k)} = \log S - T$$

where $S = N^{-1}\sum_{n=0}^{N-1} X_n$ and $T = N^{-1}\sum_{n=0}^{N-1}\log X_n$. Function $\frac{\Gamma'(k)}{\Gamma(k)}$ is known as the "digamma" function, denoted as ψ. We have:

$$k - \psi(k) = \log S - T \qquad [5.10]$$

The solution to [5.10] has no simple analytic expression, but may be calculated using a numerical procedure such as the Newton-Raphson algorithm.

5.2.25.– (Singularity in the MLE approach) (see p. 90) We can assume, without loss of generality, that the observations verify $x_n \neq x_0$ for any $n \geq 1$. The log-likelihood is expressed as $\ell(\theta) = \sum_{n=0}^{N-1}\log p_{X_n}(x_n;\theta)$. Choosing $\widehat{m}_0 = x_0$, the log-likelihood is written as:

$$\ell(\theta) = \log\left(\frac{1}{2\sqrt{2\pi}\sigma_0} + \frac{1}{2\sqrt{2\pi}\sigma_1}e^{-(x_0-m_1)^2/2\sigma_1^2}\right)$$

$$+ \sum_{n=1}^{N-1}\log\left(\frac{1}{2\sqrt{2\pi}\sigma_0}e^{-(x_n-x_0)^2/2\sigma_0^2} + \frac{1}{2\sqrt{2\pi}\sigma_1}e^{-(x_n-m_1)^2/2\sigma_1^2}\right)$$

Choosing any $m_1 \in \mathbb{R}$ and any $\sigma_1 \in \mathbb{R}^+$, let us make σ_0 tend toward 0. The first term goes to $+\infty$ and the second term, as $x_n - x_0 \neq 0$, tends toward a finite value. Therefore, ℓ tends toward infinity. This means that the maximum of ℓ as a function of θ is infinite. To avoid this situation, a constraint must be introduced:

– either by requiring the variances to be above a certain threshold, which is equivalent to restricting the domain by replacing Θ with $\widetilde{\Theta} \subset \Theta$, and writing $\widehat{\theta} = \arg\max_{\widetilde{\Theta}} \ell(\theta)$,

– or, using a Bayesian approach, considering a probability distribution for θ with density $f(\theta)$ and writing:

$$\widehat{\theta} = \arg\max_{\Theta} \ell(\theta) + \log f(\theta)$$

– or by imposing any other form of constraint as, for example, computing the means with at least two unequal observations.

5.2.26.– (Parameters of a homogeneous Markov chain) (see p. 90)

1) Using the Bayes rule and the Markov property, we have:

$$\mathcal{L} = \mathbb{P}\{X_0 = x_0, \ldots, X_{N-1} = x_{N-1}\}$$
$$= \mathbb{P}\{X_{N-1} = x_{N-1} | X_{N-2} = x_{N-2}, \ldots, X_0 = x_0\}$$
$$\mathbb{P}\{X_{N-2} = x_{N-2}, \ldots, X_0 = x_0\}$$
$$= \mathbb{P}\{X_{N-1} = x_{N-1} | X_{N-2} = x_{N-2}\} \mathbb{P}\{X_{N-2} = x_{N-2}, \ldots, X_0 = x_0\}$$

Iterating, we obtain:

$$\mathcal{L} = \sum_{n=1}^{N-1}\sum_{s=0}^{S-1}\sum_{s'=0}^{S-1} p_{s|s'} \mathbb{1}(x_n = s, x_{n-1} = s') \sum_{s=0}^{S-1} \alpha_s \mathbb{1}(x_0 = s)$$

Taking the log, we have:

$$\log \mathbb{P}\{X_0 = x_0, \cdots, X_{N-1} = x_{N-1}\} =$$
$$\sum_{n=1}^{N-1}\sum_{s=0}^{S-1}\sum_{s'=0}^{S-1} \log p_{s|s'} \mathbb{1}(x_n = s, x_{n-1} = s') + \sum_{s=0}^{S-1} \log \alpha_s \mathbb{1}(x_0 = s)$$

2) The estimator of the maximum likelihood of α_s is the solution to

$$\{\{1.6 \quad \begin{array}{l} \max \sum_{s=0}^{S-1} \log \alpha_s \mathbb{1}(x_1 = s) \\ \sum_s \alpha_s = 1 \end{array}$$

Hence $\alpha_s = \sum_{s=0}^{S-1} \mathbb{1}(x_0 = s)$. To clarify, the values of α_s are all null, except for that for which the position is the observed value of X.

The estimator of the maximum likelihood of $p_{s|s'}$ is the solution to

$$\begin{cases} \max \sum_{n=1}^{N-1} \sum_{s=0}^{S-1} \sum_{s'=0}^{S-1} \log p_{s|s'} \mathbb{1}(x_n = s, x_{n-1} = s') \\ \forall s', \sum_{s=0}^{S-1} p_{s|s'} = 1 \end{cases}$$

The Lagrangian is written as $\sum_{n=1}^{N-1} \sum_{s=0}^{S-1} \sum_{s'=0}^{S-1} \log p_{s|s'} \mathbb{1}(x_n = s, x_{n-1} = s') + \sum_{s'=0}^{S-1} \lambda_{s'}(\sum_{s=0}^{S-1} p_{s|s'} - 1)$. Canceling the derivative of the Lagrangian in relation to $p_{s|s'}$, we obtain:

$$\frac{1}{p_{s|s'}} \sum_{n=1}^{N-1} \mathbb{1}(x_n = s, x_{n-1} = s') + \lambda_{s'} = 0$$

and

$$\widehat{p}_{s|s'} = \frac{\sum_{n=1}^{N-1} \mathbb{1}(x_n = s, x_{n-1} = s')}{\sum_{s=0}^{S-1} \sum_{n=1}^{N-1} \mathbb{1}(x_n = s, x_{n-1} = s')}$$

The meaning of this expression is evident: we count the number of ordered pairs of form (s', s) and divide the result by the total number of pairs beginning with s'. In practice, we simply need to calculate the numerator for every (s', s) and then apply $\sum_{s=1}^{S} \widehat{p}_{s|s'} = 1$.

3) Type the program:

```
# -*- coding: utf-8 -*-
"""
Created on Sun Jun 19 08:05:14 2016
****** estimMarkovchain
@author: maurice
"""
from numpy.random import rand
from numpy import sum, ones, cumsum, zeros
from numpy import matrix as mat
S = 3; Prand = rand(S,S);
mataux = (mat(ones([S,1])) * mat(ones([1,S])))*Prand
P = Prand / mataux; alpha = rand(S); alpha = alpha/sum(alpha);
cumsumalpha = cumsum(alpha); cumsumP = cumsum(P, axis=0);
N = 10000; X = zeros(N); U = rand(); X[0] = S-sum(cumsumalpha>=U);
for n in range(1,N):
    ind=int(X[n-1]); Pn=cumsumP[:,ind]; U = rand();X[n] = S-sum
    (Pn>=U)
hatP = zeros([S,S]);
```

```
for n in range(1,N):
    ind1 = int(X[n]); ind2=int(X[n-1]); hatP[ind1,ind2] = hatP
    [ind1,ind2]+1;
mataux = (mat(ones([S,1]))) * mat(ones([1,S])))*hatP
hatP = hatP / mataux; print(P) ; print('*****') ; print(hatP)
```

5.2.27.– **(GMM generation)** (see p. 97)

– Type the following module `toolGMM`, which will also be used in exercise 5.2.28:

```
# -*- coding: utf-8 -*-
"""
Created on Thu Jun  2 07:58:07 2016
****** toolGMM
@author: maurice
"""
from numpy import zeros, sqrt, sum, size, inf
from numpy.random import rand, randn
from numpy import log, sort, ones, pi, mean, std, abs, dot, exp
def geneGMM(N,alphas,mus,sigma2s):
    """
    Generate a mixture of K gaussians
    SYNOPSIS
        [x,states]=GENEGMM(N,alphas,mus,sigma2s)
    N       = length of the sequence
    alphas  = ponderation array (K x 1)
    mus     = mean array (K x 1)
    sigma2s = variance array (K x 1)
    x       = data array N x 1
    states  = state array N x 1
    """
    sigma = sqrt(sigma2s);
    K = len(mus);
    cumalphas = zeros(K+1);
    for ii in range(K):
        cumalphas[ii+1] = cumalphas[ii]+alphas[ii];
    x = zeros(N); states = zeros(N);
    for ii in range(N):
        mm = K-sum(cumalphas>rand());
        states[ii] = mm;
        x[ii] = mus[mm]+ sigma[mm]*randn();
    return x,states
#========================================
def estimState_GMM_EM(y,alphas,mus,sigma2s):
```

```
    """
    EM algorithm for GMM
    with S components and dimension D
    SYNOPSIS
        [S,PS]=estimState_GMM_EM(y,alphas,mus,sigma2s)
    Inputs
        y        = data (Nx1)
        alphas   = (Kx1) vector of proportion
        mus      = (Kx1) vector of means
        sigma2s  = (Kx1) vector of variances
    Outputs
        S  = Nx1 state sequence, valued in (1,...,K)
        PS = probability of S
    """
    logfact2pi = log(sqrt(2*pi));
    K = len(alphas);
    N = len(y);
    logdet     = log(sigma2s);
    bkn        = zeros([K,N]);
    for k in range(K):
        difference = y-mus[k];
        aux1       = (difference **2) / sigma2s[k];
        logbns     = -0.5*(aux1+logdet[k])-logfact2pi;
        bkn[k,:]   = exp(logbns)*alphas[k];
    cnp    = sum(bkn,axis=0);
    gamma  = bkn / dot(ones([K,1]),cnp.reshape(1,N));
    # [PS, S] = max(gamma,[],1);
    PS = gamma.max(axis=0);
    S = gamma.argmax(axis=0)
    return S, PS
#==========================================
def estimparamGMM_EM(y,K,tol=1e-5,ITERMAX=1000):
    """
    EM algorithm for GMM with S components and
    dimension D
    SYNOPSIS
        [alphas,mus,sigma2s,llike]=...
        ESTIMGMM_EM(y,K,tol,ITERMAX)
    Inputs
        y       = data N x 1
        K       =
        tol     = tolerance
        ITERMAX = maximal number of iterations
```

```
Outputs
   alphas  = final proportion
   mus     = final means
   sigma2s = final variances
   llike   = log-likelihood
"""
N = size(y,0);
llike = zeros(ITERMAX);
likely_old = -inf; relativediff = +inf;
kEM = 0; logfact2pi = log(sqrt(2*pi));
# initialization
alphas = ones(K)/float(K);
ysort = sort(y);
NK = int(N/K)*K;
ysort = ysort[range(NK)]
yK = ysort.reshape(K,int(NK/K));
mus = mean(yK,axis=1); sigma2s = 2.0*std(yK,axis=1)**2;
while (relativediff>tol) & (kEM<ITERMAX):
    logdet = log(sigma2s);
    bkn = zeros([K,N]);
    for k in range(K):
        difference = y-mus[k];
        aux1 = (difference **2) / sigma2s[k];
        logbns = -0.5*(aux1+logdet[k])-logfact2pi;
        bkn[k,:] = exp(logbns)*alphas[k];
    cnp = sum(bkn,axis=0);
    gamma = bkn / dot(ones([K,1]),cnp.reshape(1,N));
    llike[kEM] = sum(log(cnp));
    sum_gamma_nn = sum(gamma,axis=1);
    alphas = sum_gamma_nn/float(N);
    #===== mean re-estimation
    mus = dot(gamma,y.reshape(N)) / sum_gamma_nn.reshape(K);
    #===== variance re-estimation
    aux20 = dot(y.reshape(N,1),ones(K).reshape(1,K))
    aux21 = dot(ones([N,1]),mus.reshape(1,K))
    aux2 = (aux20-aux21)**2;
    aux3 = gamma * aux2.transpose();
    sigma2s = sum(aux3,axis=1) / sum_gamma_nn;
    relativediff = abs((llike[kEM]/likely_old)-1.0);
    likely_old = llike[kEM];
    kEM = kEM+1;
llike = llike[range(1,kEM)];
alphas = sort(alphas)
```

```
        mus = sort(mus)
        sigma2s = sort(sigma2s)
        return alphas,mus,sigma2s,llike
```

– Type the program:

```
# -*- coding: utf-8 -*-
"""
Created on Thu Jun  2 08:41:35 2016
****** estimparamGMM
@author: maurice
"""
from toolGMM import geneGMM, estimparamGMM_EM
from numpy import array, sort
from matplotlib import pyplot as plt
alphas=array([0.2,0.3,0.1,0.4])
N=2000; mus = array([1.0,-1.0,3.0,-2.0]);
sigma2s=array([0.1,0.2,0.1,0.4])
aux = geneGMM(N,alphas,mus,sigma2s)
y = aux[0]; N = len(y); K=4;
alphashat,mushat,sigma2shat,llike = \
    estimparamGMM_EM(y,K,tol=1e-6,ITERMAX=1000)
print('***** ')
print('***** alpha')
alphassort = sort(alphas)
print('%4.2f,%4.2f,%4.2f,%4.2f'%(alphassort[0],alphassort[1],\
      alphassort[2],alphassort[3]))
print('%4.2f,%4.2f,%4.2f,%4.2f'%(alphashat[0],alphashat[1],\
      alphashat[2],alphashat[3]))
print('***** mu')
mussort = sort(mus); mushatsort = sort(mushat);
print('%4.2f,%4.2f,%4.2f,%4.2f'%(mussort[0],mussort[1],\
      mussort[2],mussort[3]))
print('%4.2f,%4.2f,%4.2f,%4.2f'%(mushatsort[0],mushatsort[1],\
      mushatsort[2],mushatsort[3]))
print('***** sigma2')
sigma2sort = sort(sigma2s); sigma2shatsort = sort(sigma2shat);
print('%4.2f,%4.2f,%4.2f,%4.2f'%(sigma2shat[0],\
      sigma2sort[1],sigma2sort[2],sigma2sort[3]))
print('%4.2f,%4.2f,%4.2f,%4.2f'%(sigma2shat[0],\
      sigma2shatsort[1],sigma2shatsort[2],sigma2shatsort[3]))
plt.clf(); plt.plot(llike,'.-r'); plt.show()
```

5.2.28.– (Estimation of states of a GMM) (see p. 97)

1) Equation [2.128] gives the probability of state S_n at instant n, conditional on the observation Y_n. Knowing parameter θ, we can therefore use the EM algorithm to calculate:

$$\widehat{S} = \arg\max_{k \in \{1, \ldots, K\}} \mathbb{P}_\theta \{S = k | Y\}.$$

2) Type the program:

```
# -*- coding: utf-8 -*-
"""
Created on Thu Jun  2 14:59:42 2016
****** estimStateGMM
@author: maurice
"""
from toolGMM import geneGMM, estimState_GMM_EM, estimparamGMM_EM
mus= [1,5,8]; sigma2s = [1.0,0.5,1.2]; alphas = [0.5,0.3,0.2];
K = len(alphas); Nlearn = 10000;
xlearn=geneGMM(Nlearn,alphas,mus,sigma2s); tol=1e-8; ITERMAX=180;
#===== learning
[hatalphas,hatmus,hatsigma2s,llike]= \
        estimparamGMM_EM(xlearn[0],K,tol,ITERMAX);
#===== testing data
Ntest = 400; xtest = geneGMM(Ntest,hatalphas,hatmus,hatsigma2s);
Strue = xtest[1];
[hatS,PS]=estimState_GMM_EM(xtest[0],hatalphas,hatmus,hatsigma2s);
errorrate = 1.0-sum((Strue==hatS))/float(Ntest);
print('Error rate %4.4f'%errorrate)
```

5.2.29.– (MLE on censored data) (see p. 99)

1) We have $F_X(x;\theta) = 1 - e^{-\theta x}\mathbb{1}(x \geq 0)$. Using expression [2.132], we have:

$$Q(\theta, \theta') = N\log(\theta) - \theta \sum_{n=0}^{N-1} Y_n - \frac{\theta}{\theta'} \sum_{n=0}^{N-1} \mathbb{1}(c_n = 1)$$

2) Canceling the derivative with respect to θ, we obtain the recursion of the EM algorithm:

$$\frac{1}{\theta^{(p)}} = m_Y + \rho_1 \frac{1}{\theta^{(p-1)}}$$

where $\rho_1 = N^{-1}\sum_{n=0}^{N-1} \mathbb{1}(c_n = 1)$ and $m_Y = N^{-1}\sum_{n=0}^{N-1} Y_n$. This equation converges, because $\rho_1 < 1$. The limit therefore verifies the equation, giving:

$$\theta_{\lim} = \frac{1 - \rho_1}{m_Y} \qquad [5.11]$$

Using expression [2.130], we obtain a likelihood expressed as $\mathcal{L}_{\lim} = N_0 \log N_0/S - N_0$

3) Program censoredsimul.py carries out a simulation with $N = 30$. A total of 10 censored data points are obtained by random reduction of the "true" values. We see that the mean square error is better for the estimator [5.11] than for the estimator that considers that none of the data points are censored, and than for the estimator that only takes account of the uncensored data.

```
# -*- coding: utf-8 -*-
"""
Created on Wed Jun  1 09:35:41 2016
***** censoredsimul
@author: maurice
"""
from numpy import zeros, mean, log, std
from numpy.random import rand
theta0=2; N=20; nc=10;
Lruns=300; tt=zeros([Lruns,3]);
for ir in range(Lruns):
    Y = -log(rand(N))/theta0; c = zeros(N); c[range(nc)]=1;
    Y[range(nc)] = 0.2*(rand(nc) * Y[range(nc)]);
    meany = mean(Y); rho0 = mean(c==0); tt[ir,0] = rho0/meany;
    tt[ir,1] = 1/mean(Y); tt[ir,2] = 1/mean(Y[nc:N]);
print(std(tt-theta0,axis=0))
```

4) Program censoredHT.py estimates the expectation of life after a heard implant.

```
# -*- coding: utf-8 -*-
"""
Created on Wed Jun  1 09:15:12 2016
****** censoredHT
@author: maurice
"""
from numpy import mean
import statsmodels.api as sm
heart = sm.datasets.heart.load()
survival = heart.endog/365.0; c = heart.censors
```

```
meany = mean(survival); rho0 = mean(c==0); thetalim = rho0/meany
print('***** the life expectation is %4.2f years'%(1.0/thetalim))
```

5.2.30.– (Heart implant, model with exogenous variable) (see p. 100) Starting with the probability density of the non-censored data:

$$\log p(x; \theta(a)) = \log \theta(a) - \theta(a)x$$

the expression [2.132] writes:

$$Q(\theta, \theta') = \sum_{n=0}^{N-1} \log(\alpha_0 + \alpha_1 a_n) - \sum_{n=0}^{N-1} Y_n(\alpha_0 + \alpha_1 a_n)$$

$$- \sum_{n=1}^{N} \mathbb{1}(c_n = 1) \frac{\alpha_0 + \alpha_1 a_n}{\alpha_0' + \alpha_1' a_n}$$

Maximization with respect to α_0 and α_1 is given by canceling the respective derivatives:

$$\partial_{\alpha_0} Q(\theta, \theta') = \sum_{n=0}^{N-1} \frac{1}{\alpha_0 + \alpha_1 a_n} - \sum_{n=0}^{N-1} Y_n - \sum_{n=0}^{N-1} \mathbb{1}(c_n = 1) \frac{1}{\alpha_0' + \alpha_1' a_n} = 0$$

$$\partial_{\alpha_1} Q(\theta, \theta') = \sum_{n=0}^{N-1} \frac{a_n}{\alpha_0 + \alpha_1 a_n} - \sum_{n=0}^{N-1} Y_n a_n - \sum_{n=0}^{N-1} \mathbb{1}(c_n = 1) \frac{a_n}{\alpha_0' + \alpha_1' a_n} = 0$$

This can be rewritten as:

$$\sum_{n=0}^{N-1} \frac{1}{\alpha_0 + \alpha_1 a_n} = A, \text{ where } A = \sum_{n=0}^{N-1} Y_n + \sum_{n=1}^{N} \mathbb{1}(c_n = 1) \frac{1}{\alpha_0' + \alpha_1' a_n} \quad [5.12]$$

$$\sum_{n=0}^{N-1} \frac{a_n}{\alpha_0 + \alpha_1 a_n} = B, \text{ where } B = \sum_{n=0}^{N-1} Y_n a_n + \sum_{n=0}^{N-1} \mathbb{1}(c_n = 1) \frac{a_n}{\alpha_0' + \alpha_1' a_n}$$

Let us note that $\alpha_0 A + \alpha_1 B = N$; therefore, $\alpha_0 = (N - \alpha_1 B)/A$. Carrying this expression in [5.12], we obtain:

$$f(\alpha_1) = \sum_{n=0}^{N-1} \frac{1}{\alpha_1(a_n - B/A) + N/A} - A \quad [5.13]$$

and look for the zero-crossing of f. We can note the Hessian is always negative, leading to a unique maximum. Therefore, the zero-crossing exists if the derivative in

0 is negative, i.e. $\sum_n a_n < NB/A$. Also, very large f values occur when the denominators are close to 0. Then, if $\sum_n a_n < NB/A$, we can restrict the range of value of α_1 to $-N/(A\min a_n - B)$. For the recorded value of ages, we find that $\alpha_1 = 0$. Try the random age values proposed in the following program:

```
# -*- coding: utf-8 -*-
"""
Created on Sat Jun 25 05:48:11 2016
****** censoredHTwithexogenous
@author: maurice
"""
from numpy import mean, sum, zeros, linspace, log, exp
import statsmodels.api as sm
from matplotlib import pyplot as plt
heart = sm.datasets.heart.load()
survival_year = heart.endog/365.0
age = heart.exog; c = heart.censors; n = len(c)
# uncomment to test with random age values
# ss=8939; seed(ss); age = 50.0*rand(n)
sumY = sum(survival_year); sumYa = sum(survival_year*age)
meanY = sumY/n; sumcensored = sum(c==1)
alpha0 = (1.0-sumcensored/float(n))/mean(survival_year)
alpha1 = 1.0; alpha0prime = alpha0; alpha1prime = alpha1
thetanprime = alpha0prime+alpha1prime*
age
La1=1000; f = zeros(La1); Lruns = 100; loglikely=zeros(Lruns)
for it in range(Lruns):
    sumcensoredprime = sum( (c==1)/(alpha0prime+alpha1prime*age))
    A = sumY+sumcensoredprime
    sumcensoredaprime=sum((c==1)*(age/(alpha0prime+alpha1prime*age)))
    B = sumYa+sumcensoredaprime
    if sum(age)>n*B/A:
        listmin= -float(n)/(A*age.min()-B)
        gamman = age-B/A; gamma0=n/A; lista1 = linspace(0.0001,
        listmin,La1)
        for ia1 in range(La1):
            a1=lista1[ia1]; f[ia1]=(sum(1.0/(a1*gamman+gamma0)))-A)
        alpha1prime = lista1[abs(f).argmin()]; alpha0prime =
        (n-alpha1prime*B)/A
        thetanprime = alpha0prime+alpha1prime*age
    else:
        alpha1prime = 0.0; alpha0prime = float(n)/A
```

```
        logp0 = sum((log(thetanprime)-sum(survival_year*thetanprime))*
        (c==0))
        logp1 = sum(log(1.0-exp(-thetanprime))*(c==1)); loglikely[it]
        =logp0+logp1
plt.clf(); plt.subplot(211); plt.plot(f,'.-')
plt.ylim([-10,100]); plt.subplot(212); plt.plot(loglikely,'.-')
plt.title('$\\alpha_0$ = %4.2e, $\\alpha_1$ = %4.2e'\
        %(alpha0prime, alpha1prime))
plt.show()
```

5.2.31.– **(Logistic regression)** (see p. 102) The following function `toollogisticNR` of the module `toollogistic` estimates the parameters of a logistic model, and is based on the Newton-Raphson algorithm.

```
# -*- coding: utf-8 -*-
"""
Created on Wed Jun  8 16:10:06 2016
****** toollogistic
@author: maurice
"""
from numpy import size, inf, exp, array, log, zeros
from numpy.linalg import inv
from scipy import cosh
from numpy import matrix as mat
#=============================================
def logisticNR(Z,S,tol,ITERMAX):
    """
    Newton-Raphson algorithm
    SYNOPSIS
        logisticNR(Z,S,tol,ITERMAX)
    Inputs:
        Z=array Nx(p+1),explicative variables (first column of 1s)
        S = array N x 1, response in {0,1}
        tol = relative gap, typically 1e-8
        ITERMAX = maximal iteration number, typically 100
    Outputs:
        alpha = regression coefficients (p+1) x 1
        loglike = log-likelihood
        Covalpha = covariance of the alpha estimates
        likp = 1/(1+exp(Z*alpha))
    """
    p = size(Z,1); N = size(Z,0);
    Zmat = mat(Z)
    sumZS = sum(Zmat[S==1,:], 0)
```

```
relativediff = +inf; ell_old = -inf;
alpha = zeros(p);# alpha[0]=1;
alpha = mat(alpha)
alpha = alpha.transpose()
loglike = zeros(ITERMAX);
kNR = 0;
while (relativediff>tol) & (kNR<ITERMAX):
    linkp = 1.0 / (1.0+exp(Zmat*alpha));
    aux2 = linkp.transpose()*Zmat
    dell = -sumZS+aux2;
    d2ell = zeros([p,p]);
    for nn in range(N):
        aux31 = Zmat[nn,:]*alpha
        aux3 = exp(aux31[0,0])/(1.0+exp(aux31[0,0]))**2
        d2ell = d2ell + aux3 * (Zmat[nn,:].transpose()*
        Zmat[nn,:]);
    invd2ell = inv(d2ell)
    alpha = alpha + invd2ell*dell.transpose();
    aux = Zmat*alpha
    loglike[kNR] = sum(aux[S==0])-\
                    sum(log(1.0+exp(Zmat*alpha)));
    relativediff = abs(loglike[kNR]/ell_old-1);
    ell_old = loglike[kNR];
    ell_old
    kNR = kNR+1;
loglike = loglike[1:kNR];
aux4 = 1.0 / (2.0*cosh(Zmat*alpha/2.0))
Ztilde = mat(zeros([N,p]))
for ip in range(p):
    Ztilde[:,ip]=array(Zmat[:,ip])*array(aux4)
Covalpha = inv(Ztilde.transpose()*Ztilde);
return [alpha, loglike, Covalpha, linkp]
```

The program logisticORing.py estimates the parameters of the data provided in Table 2.4, along with the associated confidence intervals, given by expression [2.138]. It calculates the p-value of the log-GLRT, expression [2.29], associated with the hypothesis $H_0 = \{\alpha_2 = 0\}$.

We see that the p-value of the log-GLRT associated with hypothesis H_0 is 0.01, leading us to reject H_0. Furthermore, the confidence interval at 95%, $0.01 \leq \alpha_2 \leq 0.34$, does not contain the value 0, which also leads us to reject H_0.

Type the following program:

```python
# -*- coding: utf-8 -*-
"""
Created on Fri Jun  3 23:28:15 2016
****** logisticORing
@author: maurice
"""
from numpy import array,ones, zeros
from numpy import diag, sqrt
from scipy.stats import chi2
from toollogistic import logisticNR
dataK = ([53,56,57,63,66,67,67,67,68,69,70,70,70,70,72,73,
75,75,75,76,78,79,80,81]);
dataS = ([1,1,1,0,0,0,0,0,0,0,0,1,1,1,0,0,0,1,0,0,0,0,0,0]);
LK = len(dataK); Z = zeros([LK,2]); Z[:,0] = ones(LK);
Z[:,1] = array(dataK,dtype='float') #dataK;
S = array(dataS); tol=1e-10; ITERMAX = 100;
result = logisticNR(Z,S,tol,ITERMAX);
alpha = result[0]; ell = result[1]; Covalpha = result[2]; Nell=len
(ell)-1
#===== test H0 = {alpha[1]=0}
result0 = logisticNR(ones([LK,1]),S,tol,ITERMAX);
alpha0 = result0[0]; ell0 = result0[1]; Covalpha0 = result0[2]
Nell0 = len(ell0)-1; T = 2.0*(ell[Nell]-ell0[Nell0]); pvalue =
1-chi2.cdf(T,1);
if pvalue<0.05:
    decision='H0 false'
else:
    decision='H0 true'
DC = diag(Covalpha);
Ib0i = alpha[0] - 1.96*sqrt(DC[0]); Ib0s = alpha[0] + 1.96*sqrt
(DC[0]);
Ib1i = alpha[1] - 1.96*sqrt(DC[1]); Ib1s = alpha[1] + 1.96*sqrt
(DC[1]);
print('****************')
print('\t%5.2f \t< alpha[0] =%5.2f < %5.2f'\
        %(Ib0i,alpha[0],Ib0s))
print('\t%5.2f \t< alpha[1] =%5.2f  < %5.2f'\
        %(Ib1i,alpha[1],Ib1s))
print('\tp-value of H0={alpha2=0} = %4.2f==>%s'%(pvalue,decision))
```

5.2.32.– (GLRT for the logistic model) (see p. 102) The program logistictestGLRT.py verifies the distribution of the GLRT under H_0. It uses the function logisticNR proposed in exercise 5.2.31.

Type the following program:

```
# -*- coding: utf-8 -*-
"""
Created on Wed Jun  8 09:22:45 2016
****** logistictestGLRT
@author: maurice
"""
from numpy import ones, dot, array, exp, zeros,linspace
from numpy.random import randn, rand
from scipy.stats import chi2
from matplotlib import pyplot as plt
from toollogistic import logisticNR
#======================================
tol=1e-8; ITERMAX = 100; N = 100;
alpha_true0 = array([0.3,0.5,1.0,0.0,0.0,0.0]); p = len(alpha_true0);
#===== design matrix randomly selected
X = randn(N,p-1); Z = zeros([N,p]); Z[:,0] = ones(N); Z[:,range(1,p)] = X
r = 1; Z0 = zeros([N,p-r]); Z0[:,0] = ones(N);
Z0[:,range(1,p-r)] = X[:,range(p-r-1)]
Za = dot(Z,alpha_true0); proba1 = 1.0 / (1.0+exp(Za));
Lruns = 500; GLRT = zeros(Lruns);
for irun in range(Lruns):
    Y = rand(N) < proba1; results1 = logisticNR(Z,Y,tol,ITERMAX);
    alphaaux1 = results1[0]; ell1 = results1[1]
    C1 = results1[2]; results0 = logisticNR(Z0,Y,tol,ITERMAX);
    alphaaux0 = results0[0]; ell0 = results0[1]
    C0 = results0[2]; GLRT[irun] = 2.0*(ell1[len(ell1)-1]-ell0[len(ell0)-1]);
bins=50;
plt.clf(); auxhist = plt.hist(GLRT,bins,normed='True');
plt.hold('on'); xtheo = linspace(0,10,200)
plt.plot(xtheo,chi2.pdf(xtheo,r),'-r');plt.hold('off');plt.show()
```

5.2.33.– (Logistic regression on home owner/rent) (see p. 102) Type and run the following program:

```
# -*- coding: utf-8 -*-
```

```
"""
Created on Fri Jun  3 23:28:15 2016
****** logisticOwner
@author: maurice
"""
from numpy import ones, zeros
from numpy import diag, sqrt
from scipy.stats import chi2
from toollogistic import logisticNR
import statsmodels.api as sm
data = sm.datasets.ccard.load(); age = data.data['AGE']
income = data.data['INCOME']; S = data.data['OWNRENT']
N = len(S); p = 2; Z = zeros([N,p+1]); Z[:,0] = ones(N);
Z[:,1] = age ; Z[:,2] = income; tol=1e-10; ITERMAX = 100;
result = logisticNR(Z,S,tol,ITERMAX);
alpha = result[0]; ell = result[1]; Covalpha = result[2]
Nell=len(ell)-1
#===== under H0={alpha[1]=alpha[2]=0}
result0 = logisticNR(ones([N,1]),S,tol,ITERMAX);
alpha0 = result0[0]; ell0 = result0[1]; Covalpha0 = result0[2]
Nell0 = len(ell0)-1
T = 2.0*(ell[Nell]-ell0[Nell0]); pvalue = 1-chi2.cdf(T,1);
if pvalue<0.05:
    decision='H0 false'
else:
    decision='H0 true'
DC = diag(Covalpha); Ibi = zeros(p+1); Ibs = zeros(p+1)
for ip in range(p+1):
    Ibi[ip] = alpha[ip] - 1.96*sqrt(DC[ip]);
    Ibs[ip] = alpha[ip] + 1.96*sqrt(DC[ip]);
print('****************')
for ip in range(p+1):
    print('\t%5.2f \t< alpha[%i] =%5.2f < %5.2f'\
        %(Ibi[ip],ip,alpha[ip],Ibs[ip]))
print('\tp-value of   H0 = {alpha[1]=alpha[2]=0} = %4.2e\
  ==> %s'%(pvalue,decision))
```

5.2.34.– **(Cumulative function estimation)** (see p. 106) Type and run the following program:

```
# -*- coding: utf-8 -*-
"""
Created on Sun May 29 17:56:03 2016
****** cumulEstimate
```

```
@author: maurice
"""
from scipy.stats import norm
from numpy.random import rand, randn
from numpy import cumsum, sort, zeros, arange
from matplotlib import pyplot as plt
N=1000; XGauss=randn(N); XGausssort = sort(XGauss);
Xmulti = zeros(N); pi0 = [0.5, 0.25, 0.125, 0.125];
Lp = len(pi0); trueCF = cumsum(pi0);
for n in range(N):
    Xmulti[n] = sum(trueCF<rand())
Xmultisort = sort(Xmulti); rangeY=arange(float(N))/N
plt.clf(); plt.subplot(211)
plt.plot(XGausssort,arange(float(N))/N,'r'); plt.hold('on')
plt.plot(XGausssort,norm.cdf(XGausssort,0.0,1.0))
plt.hold('off'); plt.title('Gaussian case')
plt.subplot(212); plt.plot(Xmultisort,rangeY); plt.hold('on')
plt.plot(range(Lp),trueCF,'o')
plt.hold('off'); plt.xlim([0,3.1]); plt.ylim([0,1.1])
plt.xticks(range(Lp)); plt.grid('on')
plt.title('Multinomial case'); plt.show()
```

5.2.35.– **(Estimation of a quantile)** (see p. 106)

1) An estimator is given by:

$$\widehat{s}_N = X_{(\lfloor cN \rfloor)}$$

where $X_{(n)}$ is the nth value of the series arranged in increasing order, known as the order statistic. This estimator may be refined by approximating the cumulative function locally around $X_{(\lfloor cN \rfloor)}$ by a polynomial.

2) Applying the δ-method to expression [2.141], we can deduce the asymptotic distribution of $\widehat{s}_N$:

$$\sqrt{N}(\widehat{s}_N - s) \to N(0, \eta) \text{ where } \eta = \frac{F(s)(1 - F(s))}{\left(\frac{dF}{ds}\right)^2}$$

3) From this, we deduce an approximate confidence interval at 95%:

$$I = \left(\widehat{s}_N - \frac{1.96\sqrt{\widehat{\gamma}}}{p(\widehat{s}_N)\sqrt{N}}, \widehat{s}_N + \frac{1.96\sqrt{\widehat{\gamma}}}{p(\widehat{s}_N)\sqrt{N}}\right)$$

4) Type the following program:

```python
# -*- coding: utf-8 -*-
"""
Created on Mon May 30 06:54:23 2016
******** CumulInverseEstimate
@author: maurice
"""
from numpy.random import randn
from numpy import sqrt, zeros, log, sort, dot, array, ones
from numpy.linalg import inv
from scipy.stats import norm, t
Lruns = 1000; val = zeros(Lruns); ICval = zeros(Lruns);
CIpercent = 0.95; N   = 10000;
#===== a few percent of the values are taken into
# account when calculating the value of pdf
# around the value of interest
dxis = 0.08; xinf = 1.0-dxis/2.0; xsup = 1.0+dxis/2.0;
alphapercent = 0.9; calpha = norm.isf((1.0-alphapercent)/2.0) / sqrt(N);
goodchoice = 'True'
#======== choose 'G' or 'R' or 'T'
case = 'G'
if case =='G':
    Xa = randn(N,Lruns);
    valtrue = norm.isf(1.0-CIpercent);
elif (case =='R'):
        Xa = sqrt(randn(N,Lruns)**2 + randn(N,Lruns)**2);
        valtrue = sqrt(-2*log(1.0-CIpercent));
elif (case =='T'):
        Xa = randn(N,Lruns) / randn(N,Lruns);
        valtrue = t.isf(CIpercent,1);
else:
    goodchoice == 'False'
#=========
if goodchoice:
    VV = array([1.0,CIpercent,CIpercent**2])
    indp = int(N*CIpercent);
    seqm1to1 = indp+array([-1,0,1])
    xxss = seqm1to1/float(N);
    MM = array([ones(3),xxss,xxss**2]).transpose()
    invMM = inv(MM)
    for ir in range(Lruns):
        X = Xa[:,ir]; Xsort = sort(X); yyss = Xsort[seqm1to1];
        alphass = dot(invMM,yyss); val_ir = dot(VV, alphass);
```

```
        pdf_select = sum((X> val_ir*xinf)&(X<val_ir*xsup)) \
                     /float(N)/(val_ir*dxis);
        val[ir] = val_ir;
    ICval[ir] = calpha * sqrt((1.0-CIpercent) * CIpercent)/
    pdf_select;
    outofIC = sum((val-ICval>valtrue)|(val+ICval<valtrue))
    print('* Percent of values outside : %4.1f'%(100.0*outofIC/
    float(Lruns)))
```

5.2.36.– **(Image equalization)** (see p. 106) Type and run the following program:

```
# -*- coding: utf-8 -*-
"""
Created on Wed Jun  1 21:41:46 2016
****** egalizeimage
@author: maurice
"""
from numpy import size, zeros, sum, cumsum
import scipy.misc as misc
from matplotlib import pyplot as plt
staircase = misc.ascent()
NX = size(staircase,0); NY = size(staircase,1); N = NX*NY;
hatp = zeros(256);
for it in range(256):
    hatp[it]=sum(staircase==it)/float(N);
cdfimg = cumsum(hatp); imgequal = zeros([NX,NY]);
for iX in range(NX):
    for iY in range(NY):
        itx = cdfimg[staircase[iX,iY]];
        imgequal[iX,iY] = int(255*itx);
hatpequal = zeros(256);
for it in range(256):
    hatpequal[it]=sum(imgequal==(it-1))/float(N);
cdfimgequal = cumsum(hatpequal);
plt.clf(); plt.subplot(221); plt.imshow(256-staircase,cmap='Greys');
plt.show()
plt.subplot(222); plt.imshow(256-imgequal,cmap='Greys')
plt.subplot(223); plt.plot(range(256),cdfimg); plt.grid('on')
plt.subplot(224); plt.plot(range(256),cdfimgequal); plt.grid('on')
```

5.2.37.– **(Bootstrap for a regression model)** (see p. 110) Type and run the following program:

```
# -*- coding: utf-8 -*-
```

```
"""
Created on Wed Jun  1 08:34:29 2016
****** bootstraponregression
@author: maurice
"""
from numpy import zeros,ones, dot, cov, mean
from numpy.linalg import pinv, inv
from numpy.random import randn, randint
N = 30; mu = [3,2]; p = len(mu);
Z = zeros([N,2]); Z[:,0] = ones(N); Z[:,1] = range(N)
Zmu = dot(Z,mu); B = 200; mub = zeros(B);
Lruns = 1000; sigma2b = zeros([p,p,Lruns]);
for irun in range(Lruns):
    X = Zmu + randn(N); U = randint(N,size=[N,B]); mub = zeros
    ([p,B]);
    for ib in range(B):
        ZU = Z[U[:,ib],:]; HH = pinv(ZU); mub[:,ib] =    dot(HH,X
        [U[:,ib]]);
    sigma2b[:,:,irun] = cov(mub);
theosigma2b = inv(dot(Z.transpose(),Z)); msq2b00=sigma2b[0,0,:];
m00 = mean(msq2b00);
msq2b01=sigma2b[0,1,:];m01=mean(msq2b01); msq2b11=sigma2b[1,1,:];
m11 = mean(msq2b11);
print('******* estimation of mu[0] and mu[1] ******')
print('theoretical variance on mu[0] estimate = \
%5.3e\nboot-value estimate = %5.3e\n'%(theosigma2b[0,0],m00))
print('theoretical variance on mu[1] estimate = \
%5.3e\nboot-value estimate = %5.3e\n'%(theosigma2b[1,1],m11))
print('theoretical covariance on the couple estimate = \
%5.3e\nboot-value estimate = %5.3e\n'%(theosigma2b[0,1],m01))
```

5.2.38.– **(Model selection based on cross-validation)** (see p. 112) Type and run the program orderEstimCV.py:

```
# -*- coding: utf-8 -*-
"""
Created on Fri Jun  3 07:16:04 2016
****** orderEstimCV
@author: maurice
"""
from numpy.random import randn
from numpy.linalg import norm, pinv
from numpy import ones, zeros, sum, dot, setdiff1d
from matplotlib import pyplot as plt
```

```
N = 300; K = 10; L = int(N/K); sigma = 2;
Ptrue = 10; Pmax  = 20; X = randn(N,Pmax);
Z = X[:,range(Ptrue)]; beta = ones(Ptrue);
y = sum(Z, axis=1) + sigma*randn(N);
errT = zeros([Pmax,K]); errL = zeros([Pmax,K]);
for ik in range(K):
    id1 = int(ik*L); id2 = id1+L-1;
    #===== testing DB
    Ty = y[range(id1,id2+1)]; TX = X[range(id1,id2+1),:];
    #===== learning DB
    rg = setdiff1d(range(N),range(id1,id2+1))
    Ly = y[rg]; LX = X[rg,:];
    for ip in range(1,Pmax):
        rgip = range(ip); LH = LX[:,rgip]; TH = TX[:,rgip];
        W = pinv(LH); hbeta = dot(W,Ly);
        errT[ip,ik] = norm(Ty-dot(TH,hbeta))**2 /(L-ip);
        errL[ip,ik] = norm(Ly-dot(LH,hbeta))**2 /((K-1)*L-ip);
errT_p = sum(errT,axis=1)/K;
errL_p = sum(errL,axis=1)/K;
#=====
plt.clf(); plt.plot(errL_p[1:],'x-'); plt.hold('on')
plt.plot(errT_p[1:],'o-'); plt.hold('off'); plt.grid('on');
plt.show()
```

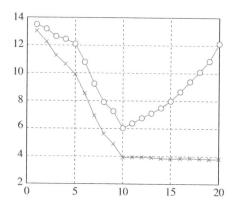

Figure 5.1. Prediction errors as a function of the supposed order of the model: 'x' for the learning base; 'o' for the test base. The observation model is of the form $y = Z\beta + \sigma\epsilon$, where Z includes $p = 10$ column vectors. As the number of predictors increases above and beyond the true value, we "learn" noise; this is known as overtraining

Figure 5.1 shows the results of a simulation. The curve marked as 'x' represents the prediction error calculated over the learning base. The mean decreases as the number of predictors increases. The curve marked as 'o', which represents the prediction error calculated using the test base, passes through a minimum for the true value $p = 10$. Thus, when the number of predictors is increased above the true value, we "learn" noise. This is known as overtraining. Furthermore, note that the error for the test base is greater than that for the learning base, as predicted by equation [2.82].

5.2.39.– (**Cross-validation on CO_2 concentration**) (see p. 112) Type the following program:

```
# -*- coding: utf-8 -*-
"""
Created on Mon Jun 20 17:18:19 2016
****** co2CV
@author: maurice
"""
from numpy import zeros, arange, pi, cos, sin
from numpy import isnan, array, sqrt, size, unravel_index
import statsmodels.api as sm
from numpy import matrix as mat
from numpy.linalg import pinv
from sklearn import cross_validation as CV
dataCO2 = sm.datasets.co2.load(); y = dataCO2.data['co2']; N =
len(y)
date = dataCO2.data['date']
# determine the nan value indices
listindex=list([])
for ip in range(N):
    if not isnan(y[ip]):
        listindex.append(ip)
t = arange(N); tprime = t[listindex]; tmat = mat(tprime).
transpose()
yprime = y[listindex]; Nmat = len(yprime); ymat = mat(yprime).
transpose()
f0 = 7.0/365.0; Lp = 7; Lq = 7; Lruns = 100; stde = array(zeros
([Lp,Lq-1]))
for q in range(1,Lq):
    for p in range(Lp):
        # learning
        for ir in range(Lruns):
            t_train, t_test, y_train, y_test = \
            CV.train_test_split(tprime,yprime,test_size=0.2)
            N_train = size(t_train,0); N_test = size(t_test,0)
```

```
            H_train = mat(zeros([N_train,2*p+q]));
            theta_train = 2*pi*f0*mat(t_train).transpose()*mat
            (arange(1,p+1))
            H_train[:,0:p] = cos(theta_train); H_train[:,p:2*p] =
            sin(theta_train)
            for iq in range(q):
                H_train[:,2*p+iq] = mat(t_train**iq).transpose()
            pinvH_train = pinv(H_train)
            alpha_train = pinvH_train*mat(y_train).transpose()
            # testing
            H_test = mat(zeros([N_test,2*p+q]))
            theta_test = 2*pi*f0*mat(t_test).transpose()*mat
            (arange(1,p+1))
            H_test[:,0:p] = cos(theta_test); H_test[:,p:2*p] =
            sin(theta_test)
            for iq in range(q):
                H_test[:,2*p+iq] = mat(t_test**iq).transpose()
            y_pred = H_test*alpha_train; e_pred = mat(y_test).
            transpose()-y_pred
            stde_pq = sqrt(e_pred.transpose()*e_pred/N_test)
            stde[p,q-1] = stde[p,q-1]+stde_pq
        stde[p,q-1] = stde[p,q-1]/float(Lruns)
pop,qop = unravel_index(stde.argmin(), stde.shape)
print('********** p = %i, q = %i'%(pop, qop))
```

5.2.40.– (**Cross-validation for home owner/rent**) (see p. 112) Type the following program:

```
# -*- coding: utf-8 -*-
"""
Created on Tue Jun 21 07:31:58 2016
****** logisticOwnerCV
@author: maurice
"""
from numpy import ones, zeros, size, exp, array, mean
from numpy import matrix as mat
from toollogistic import logisticNR
import statsmodels.api as sm
from sklearn import cross_validation as CV
data = sm.datasets.ccard.load(); age = data.data['AGE']
income = data.data['INCOME']; avgexp = data.data['AVGEXP']
incomesq = data.data['INCOMESQ']; XT=array([age,income,incomesq])
X = XT.transpose(); p = size(X,1); N = size(X,0); Y = data.data
['OWNRENT']
```

```
tol=1e-16; ITERMAX = 100; Lruns = 500; gp = zeros(Lruns); percent
_testbase = 0.2;
for ir in range(Lruns):
    X_train, X_test, y_train, y_test = \
        CV.train_test_split\
        (X,Y,test_size=percent_testbase)
    N_train = size(X_train,0); N_test = size(X_test,0)
    # learning
    Z_train = zeros([N_train,p+1]); Z_train[:,0] = ones(N_train);
    Z_train[:,1:p+1] = X_train;
    result = logisticNR(Z_train,y_train,tol,ITERMAX); alpha_train
    = result[0];
    # testing
    Z_test = zeros([N_test,p+1]); Z_test[:,0] = ones(N_test);
    Z_test[:,1:p+1] = X_test; matZ_test = mat(Z_test)
    sumZS = sum(matZ_test[y_test==1,:], 0)
    loglike = zeros(ITERMAX); linkp = 1.0 / (1.0+exp(matZ_test*
    alpha_train));
    y_pred = array(linkp>0.5, dtype='float').reshape(N_test)
    gp[ir] = sum(y_pred == y_test)/float(N_test)
print('\tGood prediction rate = %4.1f%s'%(100.0*mean(gp),'%') )
```

5.2.41.– **(Model selection with cross-validation)** (see p. 112)

Type the following program:

```
# -*- coding: utf-8 -*-
"""
Created on Wed Jul 27 11:28:56 2016
****** BSSdiabetes
@author: maurice
"""
from numpy import zeros, ones, dot, setdiff1d
from numpy.linalg import pinv
from sklearn import datasets as ds
import matplotlib.pyplot as plt
def residue_norm_CV(X,y,k):
    N,p = X.shape; r = p+1; L = int(N/k)
    rangeN = range(N); cve = 0.0;
    for ik in range(k):
        id1 = ik*L; id2 = id1+L-1;
        idTest = range(id1,id2+1); NTest = len(idTest)
        idLearn = setdiff1d(rangeN,idTest); NLearn = len(idLearn)
        ZLearn = zeros([NLearn,r]); ZLearn[:,0] = ones(NLearn);
```

```
            ZLearn[:,1:r] = X[idLearn,:]; yLearn = y[idLearn]
            pinvZLearn = pinv(ZLearn);
            hatbetaLearn = dot(pinvZLearn,yLearn);
            ZTest = zeros([NTest,r]); ZTest[:,0] = ones(NTest);
            ZTest[:,1:r] = X[idTest,:]; yTest = y[idTest]
            hatyTest = dot(ZTest,hatbetaLearn)
            eTest = (hatyTest-yTest);
            cve = cve + sum(eTest**2)
    cve = cve/NTest
    return cve
def residue_norm(X,y):
    N,p = X.shape; r = p+1;
    Z = zeros([N,r]); Z[:,0] = ones(N); Z[:,1:r] = X;
    pinvZ = pinv(Z); hatbeta = dot(pinvZ, y);
    haty = dot(Z,hatbeta); T = sum((haty-y) **2)
    return T
#========= main program ====
diabetes = ds.load_diabetes();
explanatoryVars = diabetes['data']; target = diabetes['target']
N,p = explanatoryVars.shape; X = explanatoryVars.reshape(N,p);
y = target.reshape(N,1); T_CV = zeros(p+1); kCV = 5;
T_CV[p] = residue_norm_CV(X,y,kCV); T = zeros(p+1); T[p]=residue
_norm(X,y)
suppressedVarindex = zeros(p); model_k = X; colmodel_kindex =
range(p)
for k in range(p,0,-1):
    T_k = zeros(k); rangek = range(k);
    for j in rangek:
        testmodel_jk = model_k[:,setdiff1d(rangek,j)]
        T_k[j] = residue_norm(testmodel_jk,y)
    jo = T_k.argmin()
    T[k-1] = T_k.min()
    suppressedVarindex[k-1] = colmodel_kindex[jo]
    colmodel_kindex = setdiff1d(colmodel_kindex,colmodel_
    kindex[jo])
    model_k = model_k[:,setdiff1d(rangek,jo)]
    T_CV[k-1] = residue_norm_CV(model_k,y,kCV)
plt.clf(); plt.plot(T_CV,'.-'); plt.show()
selectedcolumns = setdiff1d(range(p),suppressedVarindex[T_CV.
argmin():p])
print('***** The optimal model consists of the %i explanatory
features : %s' \
 %(len(selectedcolumns),selectedcolumns))
```

5.3. Inferences on HMM

5.3.1.– (Kalman filter derivation, scalar case) (see p. 123)

1) Using the linearity of the expectation in the evolution equation, we have:

$$X_{n+1|n} = a_n X_{n|n} + \mathbb{E}\{B_n | Y_{0:n}\}$$

Noting that B_n is independent of $Y_{0:n}$ and is centered, we deduce:

$$X_{n+1|n} = a_n X_{n|n}$$

2) If we replace Y_{n+1} with $(c_{n+1} X_{n+1} + U_{n+1})$ and use the hypothesis stating that U_{n+1} and B_{n+1} are orthogonal to $Y_0, \ldots, Y_n$, we obtain:

$$(Y_{n+1}|Y_{0:n}) = c_{n+1}(X_{n+1}|Y_{0:n})$$
$$= c_{n+1} X_{n+1|n} \qquad [5.14]$$

3) Using the property [1.34], we write:

$$X_{n+1|n+1} = (X_{n+1}|Y_{0:n}, Y_{n+1})$$
$$= (X_{n+1}|Y_{0:n}) + K_{n+1} i_{n+1}$$

where $K_{n+1} = (X_{n+1}, i_n)/(i_n, i_n)$ and $i_{n+1} = Y_{n+1} - (Y_{n+1}|Y_{0:n})$.

Using [5.14], we also get $i_n = Y_{n+1} - c_{n+1} X_{n+1|n}$. Therefore:

$$X_{n+1|n+1} = X_{n+1|n} + K_{n+1}(Y_{n+1} - c_{n+1} X_{n+1|n}) \qquad [5.15]$$

4) We have:

$$i_{n+1} = \underbrace{c_{n+1} X_{n+1} + U_{n+1}}_{Y_{n+1}} - c_{n+1} X_{n+1|n}$$
$$= c_{n+1}(X_{n+1} - X_{n+1|n}) + U_{n+1}$$

Let $P_{n+1|n} = (X_{n+1} - X_{n+1|n}, X_{n+1} - X_{n+1|n})$. Because $(X_{n+1} - X_{n+1|n})$ and U_{n+1} are orthogonal, we can write:

$$\|i_{n+1}\|^2 = \text{var}(i_{n+1}) = c_{n+1}^2 P_{n+1|n} + \sigma_U^2(n+1)$$

5) The process

$$i_n = Y_n - (Y_n|Y_{0:n-1})$$

is known as the innovation process. We verify that $\mathbb{E}\{i_n\} = 0$. By definition, i_n belongs to the linear space spanned by $Y_{0:n}$. In accordance with the projection theorem, i_{n+1} is orthogonal to the linear space generated by $Y_{0:n}$. Consequently, $i_n \perp i_{n+1}$ and, being Gaussian, they are independent.

Moreover, the linear space spanned by $Y_{0:n}$ corresponds with the linear space spanned by $i_{0:n}$. The joint distribution of $Y_{0:n}$ is therefore equal to that of $i_{0:n}$. Hence:

$$-2\log p_{Y_{0:n}}(y_{0:n}) = n\log(2\pi) + \sum_{k=0}^{n} \frac{i_k^2}{\operatorname{var}(i_k)}$$

6) Furthermore:

$$(X_{n+1}, i_{n+1}) = (X_{n+1}, X_{n+1} - X_{n+1|n})c_{n+1} + \underbrace{(X_{n+1}, U_{n+1})}_{=0}$$

$$= (X_{n+1} - X_{n+1|n}, X_{n+1} - X_{n+1|n})c_{n+1}$$

$$= P_{n+1|n}c_{n+1}$$

And hence:

$$K_{n+1} = \frac{P_{n+1|n}c_{n+1}}{\sigma_U^2(n+1) + c_{n+1}^2 P_{n+1|n}} \qquad [5.16]$$

7) Let us now determine the expression of $P_{n+1|n}$. We have:

$$X_{n+1} - X_{n+1|n} = (a_n X_n + B_n) - a_n X_{n|n}$$

This leads us to:

$$P_{n+1|n} = a_n^2 P_{n|n} + \sigma_B^2(n) \qquad [5.17]$$

stating that $X_{n|n}$ and B_n are orthogonal. Using [5.15], we obtain:

$$X_{n+1} - X_{n+1|n+1} = (X_{n+1} - X_{n+1|n}) - K_{n+1}(Y_{n+1} - c_{n+1}X_{n+1|n})$$
$$= (X_{n+1} - X_{n+1|n}) - K_{n+1}c_{n+1}(X_{n+1} - X_{n+1|n}) + K_{n+1}U_{n+1}$$

Using the fact that $X_{n+1} - X_{n+1|n}$ and U_{n+1} are orthogonal, the expression [5.16] leads to:

$$P_{n+1|n+1} = (1 - K_{n+1}c_{n+1})P_{n+1|n} \qquad [5.18]$$

If we group expressions [5.15], [5.16], [5.17] and [5.18] together, we obtain the following algorithm in accordance with the Kalman algorithm 7:

$$\begin{cases} X_{n+1|n} &= a_n X_{n|n} \\ P_{n+1|n} &= a_n^2 P_{n|n} + \sigma_B^2(n) \\ K_{n+1} &= \dfrac{P_{n+1|n} c_{n+1}}{\sigma_U^2(n+1) + c_{n+1}^2 P_{n+1|n}} \\ X_{n+1|n+1} &= X_{n+1|n} + K_{n+1}\left(Y_{n+1} - c_{n+1} X_{n+1|n}\right) \\ P_{n+1|n+1} &= (1 - K_{n+1} c_{n+1}) P_{n+1|n} \end{cases} \quad [5.19]$$

with the initial conditions $x_{0|0} = 0$ and $P_{0|0} = \mathbb{E}\left\{X_0^2\right\}$.

5.3.2.– (Denoising an AR-1 using Kalman) (see p. 124)

1) According to [3.14] and [3.15], we have:

$$K_n = \dfrac{P_{n|n-1}}{P_{n|n-1} + \sigma_u^2} \quad [5.20]$$

which leads us to $P_{n|n-1}(1 - K_n) = \sigma_u^2 K_n$. Using [3.13] and [3.18]:

$$P_{n|n-1} = a^2(1 - K_{n-1}) P_{n-1|n-2} + \sigma_b^2$$
$$= a^2 \sigma_u^2 K_{n-1} + \sigma_b^2$$

Substituting this result in expression [5.20] gives the recursive formula:

$$K_n = \dfrac{\rho + a^2 K_{n-1}}{1 + \rho + a^2 K_{n-1}} \quad [5.21]$$

Using the first step of algorithm 7, we have $P_{1|0} = a^2 \sigma_b^2/(1-a^2) + \sigma_b^2 = \sigma_b^2/(1-a^2)$, hence $K_1 = \sigma_b^2/(1-a^2)/(\sigma_b^2/(1-a^2) + \sigma_u^2) = \rho/(\rho + (1-a^2))$. From the calculation point of view, everything happens as if we started out with formula [5.21] and the initial values $\widehat{X}_{0|0} = 0$ and $K_0 = \rho/(1-a^2)$.

We can easily verify that the series K_n is an increasing monotone and bounded by 1. It therefore converges and the limit verifies the recursive equation, giving $K_{\lim} = \dfrac{-(1+\rho-a^2) + \sqrt{(1+\rho-a^2)^2 + 4\rho a^2}}{2a^2}$.

The Kalman algorithm can be summed up as follows:

1) Initial conditions: $\widehat{X}_{0|0} = 0$ and $K_0 = \rho/(1-a^2)$.

2) For $n \geq 1$:

$$\begin{cases} K_n &= \dfrac{\rho + a^2 K_{n-1}}{1 + \rho + a^2 K_{n-1}} \\ \widehat{X}_{n|n} &= a\widehat{X}_{n-1|n-1} + K_n \left(Y_n - a\widehat{X}_{n-1|n-1} \right) \end{cases}$$

2) The following program is designed to test the algorithm:

```
# -*- coding: utf-8 -*-
"""
Created on Fri Jun 10 22:15:33 2016
****** KFnoisyAR1
@author: maurice
"""
from numpy import zeros, arange
from numpy.random import randn
from scipy.signal import lfilter
from matplotlib import pyplot as plt
def KalmanFilterAR1(Y,a,rho,K0):
    T = len(Y)
    a2 = a*a;
    Kn = K0;
    xtt = zeros(T)
    for t in range(1,T):
        Kn = (rho+a2*Kn)/(1+rho+a2*Kn)
        xtt[t] = a*xtt[t-1]+Kn*(Y[t]-a*xtt[t-1])
    return xtt
#==== main program
a = 0.8; sigmaB = 0.1; sigmaU = 0.3; N = 100;
w = randn(N); alpha = -a;
x = lfilter((sigmaB,),(1.0,alpha),w);
y = x+sigmaU*randn(N);
rho = sigmaB*sigmaB/(sigmaU*sigmaU)
K0 = rho/(1.0-a**2)
xtt = KalmanFilterAR1(y,a,rho,K0);
t = arange(N)
plt.clf(); plt.plot(t,y, label='observation'); plt.hold('on');
plt.plot(t,xtt, 'x-', label='filtered');
plt.plot(t,x,'o-', label='true data'); plt.hold('off');
plt.legend(loc='best'); plt.show()
```

The results are shown in Figure 5.2.

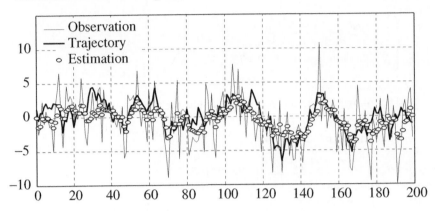

Figure 5.2. *Results for the study of the filtering*

When we presented the Kalman filter, and implemented it in the previous program, we assumed that the model as well as the characteristic features of the noise are known. However, this is usually not the case. For example, if in our case, the signal X_n is not an AR process, the choice of a and σ_b^2 requires that we compromise between the ability of X_n to track the trajectory and the elimination of the noise. Choosing a too close to 1 means that the model does not take into account the rapid variations of the signal X_n. Therefore, the filter has difficulties "keeping up" with such variations. Likewise, if we choose σ_b^2 too high, we assume that we expect significant variations of the signal X_n with respect to the equation $X_n \approx aX_{n-1}$. You can check by using the previous algorithm and changing the parameters.

5.3.3.– **(Kalman filtering of a noisy 1D trajectory)** (see p. 125) Type and run the below program. Selecting C = mat([0.0, 1.0]);, we observe that the location estimates can be very inaccurate. This is related to the aspect of non-observability of the system, as this is shown by the rank value of the matrix $\mathcal{W}$ that is different from 2.

```
# -*- coding: utf-8 -*-
"""
Created on Sat Jun 11 17:50:17 2016
****** Kalmantraj1D
@author: maurice
"""
from numpy import zeros, size, eye, array
from numpy.linalg import matrix_rank as rk
from numpy import matrix as mat
```

```python
from numpy.linalg import inv
from numpy.random import randn
from matplotlib import pyplot as plt
#================================================
def kalmanfilter(Y,A,RV,C,RW,mu0,R0):
    """
    # Kalman filter
    # SYNOPSIS
    #    [Xtt,Ptt,loglikeli]=kalmanfilter(Y,A,RV,C,RW,mu0,R0)
    # Inputs:
    #       Y    = observations (dimY x T)
    #       A    = state matrix (dimX x dimX)
    #       RV   = state covariance (dimX x dimX)
    #       RW   = observation covariance (dimY x dimY)
    #       mu0  = initial state mean (dimX x 1)
    #       R0   = initial state covariance (dimX x dimX)
    # Outputs:
    #       Xtt  = filtered state (dimX x T)
    """
    dimX            = size(A,0);
    T               = size(Y,1);
    Xtt             = mat(zeros([dimX,T]));
    Xttm1           = A * mu0;
    Rttm1           = ((A * R0) * A.transpose()) + RV;
    cov_inov        = (C * Rttm1 * C.transpose()) + RW;
    invcov_inov     = inv(cov_inov)
    Kn              = (Rttm1*C.transpose())*invcov_inov;
    inov_1          = Y[:,0]-C*Xttm1;
    Xtt[:,0]        = Xttm1+Kn*inov_1;
    Ptt             = Rttm1 - ((Kn*C)*Rttm1);
    for k in range(1,T):
        Xttm1           = A*Xtt[:,k-1];
        Rttm1           = (A*Ptt)*A.transpose() + RV;
        cov_inov        = (C*Rttm1)*C.transpose() + RW;
        invcov_inov     = inv(cov_inov)
        inov_k          = (Y[:,k] - (C*Xttm1));
        Kn              = (Rttm1*C.transpose())*invcov_inov;
        Xtt[:,k]        = Xttm1 + Kn*inov_k;
        Ptt             = Rttm1 - (Kn*C)*Rttm1;
    return Xtt
#================================================
N = 150; model1 = 1; sigmab = 1.0; sigmaU = 10.0;
A = mat([[1.0, 0.1],[0, 1.0]]); dimX = size(A,0);
```

```
D = mat([0,1]).transpose();
# select the C shape
if model1:
    C = mat([1.0, 0.0]);
else:
    C = mat([0.0, 1.0]);
W = mat(zeros([dimX,dimX])); Wik = C;
for ik in range(dimX):
    W[ik,:] = Wik
    Wik = Wik * A
print('********* rank of W is %i'%rk(W))
#== trajectory generation
RB = sigmab*sigmab*(D*D.transpose());
Xtrue = mat(zeros([dimX,N])); Xtrue[:,0] = randn(2,1)
for n in range(1,N):
    Xtrue[:,n] = A*Xtrue[:,n-1]+D*sigmab*randn()
#==== observation generation
Y = C*Xtrue + sigmaU*mat(randn(N));
#==== initial conditions
mu0 = mat([0.0,0.0]).transpose();
R0 = sigmab*sigmab*eye(2);
Xfilt = kalmanfilter(Y, A, RB, C, sigmaU**2, mu0, R0);
#====
Ya = array(Y); Xtruea = array(Xtrue); Xfilta = array(Xfilt);
plt.clf(); plt.subplot(211)
plt.plot(Xtruea[0,:],'.-g',label='true trajectory')
plt.hold('on'); plt.plot(Xfilta[0,:],'.-r',label='Kalman filter');
if model1:
    plt.plot(Ya[0,:],'.-',color=[0.6,0.6,0.6]);
plt.hold('off'); plt.legend(loc='best');
plt.subplot(212)
plt.plot(Xtruea[1,:],'.-g',label='true trajectory')
plt.hold('on')
plt.plot(Xfilta[1,1:],'.-r',label='Kalman filter');
if not(model1):
    plt.plot(Ya[0,:],'.-',color=[0.9,0.9,0.9]);
plt.hold('off');
plt.show()
```

5.3.4.– (Calculating the likelihood of an ARMA) (see p. 125)

1) The first equation of expression [3.24] is written as:

$$\begin{bmatrix} 1 & \times \\ \vdots & \vdots \\ z^{-(r-2)} & \times \\ z^{-(r-1)} & \times \end{bmatrix} \begin{bmatrix} X_{1,n} \\ \vdots \\ X_{r-1,n} \\ X_{r,n} \end{bmatrix} = \begin{bmatrix} -a_1 \\ \vdots \\ -a_{r-1} \\ -a_r \end{bmatrix} X_{1,n-1} + \begin{bmatrix} X_{n-1,2} \\ \vdots \\ X_{n-1,r} \\ 0 \end{bmatrix} + \begin{bmatrix} 1 \\ \vdots \\ b_{r-2} \\ b_{r-1} \end{bmatrix} Z_n \quad [5.22]$$

Applying a series of delays such as those given in the first column of [5.22], we obtain:

$$X_{1,n} = -\sum_{k=1}^{r} a_k X_{1,n-k} + Z_n + \sum_{m=1}^{r-1} b_m Z_{n-m}$$

Noting that, following [3.24], $X_{1,n} = Y_n$, we deduce that Y_n is an ARMA-p, q.

2) For t ranging from 2 to n, the Kalman algorithm giving the log-likelihood is written as:

$$X_{t|t-1} = A X_{t-1|t-1} \quad [5.23]$$

$$P_{t|t-1} = A P_{t-1|t-1} A^T + \sigma^2 R R^T \quad [5.24]$$

$$\gamma_t = C^T P_{t|t-1} C \quad [5.25]$$

$$K_t = \frac{1}{\gamma_t} P_{t|t-1} C \quad [5.26]$$

$$I_t = Y_t - C^T X_{t|t-1} \quad [5.27]$$

$$X_{t|t} = X_{t|t-1} + K_t I_t \quad [5.28]$$

$$P_{t|t} = P_{t|t-1} - K_t C^T P_{t|t-1} \quad [5.29]$$

$$\ell_t = \ell_{t-1} + \log(\gamma_t) + \frac{1}{\gamma_t} I_t^2 \quad [5.30]$$

with initial values

$P_{1|1}$ square matrix of size r (see remark) [5.31]

$X_{1|1} = 0$ null vector of size r [5.32]

$\ell_1 = 0$ [5.33]

3) A tilde is used to indicate quantities calculated using the algorithm with $\sigma = 1$. We consider that, on initialization, $P_{1|1} = \sigma^2 \tilde{P}_{1|1}$. We deduce that $P_{t|t-1} = $

$\sigma^2 \widetilde{P}_{T|t-1}$, $\gamma_t = \sigma^2 \widetilde{\gamma}_t$, $K_t = \widetilde{K}_t$ and $X_{t|t} = \widetilde{X}_{t|t}$. The log-likelihood is therefore written as:

$$\ell_n = \sum_{t=1}^{n} \left(\log(\sigma^2 \widetilde{\gamma}_t) + \frac{I_t^2}{\sigma^2 \widetilde{\gamma}_t} \right) \qquad [5.34]$$

We must therefore simply apply the algorithm for $\sigma = 1$ and then use formula [5.34] to calculate the likelihood. This relationship is highly useful for maximum likelihood-based estimation problems, as maximization in relation to σ^2 possesses an analytical solution.

REMARK: the choice of $P_{1|1}$ requires further explanation. According to question 3, we can consider that $\sigma^2 = 1$. One choice for $P_{1|1}$ is the matrix that corresponds to the stationary form of X_t. Noting $P_1 = \text{cov}(X_t)$ and using the evolution equation and the stationarity, we have:

$$P_1 = A P_1 A^T + Q$$

taking $Q = RR^T$. This is known as the Lyapunov equation. Using the identity:

$$\text{vec}(ABC) = (C^T \otimes A)\text{vec}(B)$$

where the operator noted vec denotes the column vectorization operation applied to a matrix. This last equation may be rewritten as:

$$\text{vec}(P_1) = (A \otimes A)\text{vec}(P_1) + \text{vec}(Q)$$

where $\otimes$ denotes the Kronecker product. Hence:

$$\text{vec}(P_1) = (I_{r^2} - A \otimes A)^{-1} \text{vec}(Q)$$

4) $a(z) X_n = b(z) Z_n$, $a(z) \widetilde{X}_n = Z_n$, then $b(z) \widetilde{X}_n = X_n$.

5) Type the following script that contains the functions LLarmaKalman.py and arma2ACF.py.

```
# -*- coding: utf-8 -*-
"""
Created on Sun Jun 12 17:57:04 2016
****** testKarmaOnARMApq
@author: maurice
"""
from numpy import zeros, dot
from numpy import log, eye, array, kron
from numpy import matrix as mat
from scipy.linalg import inv, det
```

```python
from numpy.random import randn
from scipy.signal import lfilter
from scipy.linalg import toeplitz, hankel
#==================================================
def arma2ACF(a,b,sigma2,n):
    """
    # SYNOPSIS: gamma=ARMA2ACF(a,b,sigma2,n)
    #    Compute ARMA autocovariance sequence
    # Input:
    #    a      = AR param with 1 [(p+1) x 1]
    #    b      = MA param with 1 [(q+1) x 1]
    #    sigma2 = Innovation variance
    #    n      = Number of autocovariances
    # Outputs:
    #    gamma  = Autocovariances 0 to (n-1) [n x 1]
    #    R      = Autocovariance of the AR part
    """
    p = len(a)-1;
    q = len(b)-1;
    r = max((p,q+1));
    J = zeros(p+1); J[p]=1;
    a1   = -a[1:p+1];
    #===== first p elements of AR autocovariance
    # with sigma2 = 1
    A1 = toeplitz(eye(p+1,1), a);
    A2 = hankel(a[p]*J,a[range(p,-1,-1)]);
    A = A1+A2; invA  = inv(A)
    R0 = dot(invA,J);
    R0 = R0[range(p,-1,-1)]; R0[0] = 2.0*R0[0];
    R = zeros(n+r); R[range(p+1)] = R0
    #===== following elements of AR autocovariance
    # up to (n+r-1)
    for ir in range (p+1,n+r):
        R[ir] = dot(R[range(ir-1,ir-p-1,-1)],a1);
    #===== ARMA covariance
    gamma = zeros(n);
    for k in range(n):
        gamma[k] = 0;
        for i in range(q+1):
            for j in range(q+1):
                gamma[k] = gamma[k] + \
                    b[i] * b[j] * R[abs(k - i + j)];
    # just multiply by sigma2
```

```
        gamma = gamma * sigma2;
        return gamma,R
#================================================
def LLarmaKalman(Y,a,b,sigma2):
    """
    # Likelihood of ARMA by Kalman filtering
    # SYNOPSIS: LLARMAKALMAN(Y,a,b,sigma2)
    # Inputs
    #      Y_n + \sum_k=1^p a_k Y_n-k =
    #                Z_n  + \sum_k=1^q b_k Z_n-k
    #      a = [1,a_1,...,a_p]
    #      b = [1,b_1,...,b_q]
    #      Z_n white gaussian noise with variance sigma2
    # Outputs
    #      L = likelihood maximized on the scale factor
    """
    T = len(Y); p = len(a)-1;
    q = len(b)-1; r = max([p,q+1]); r2 =r**2
    A = zeros([r,r]); matC = mat(eye(r,1));
    A[0:p,0] = -a[1:p+1];
    A[0:r-1,1:r] = eye(r-1);
    matA = mat(A)
    R = mat(b);
    Q = sigma2*(R.transpose()*R);
    #=====
    Lcurr = 0;
    Xhat_tt = mat(zeros([r,1]));
    # steady state
    bigA = eye(r2)-kron(A,A);
    invbigA = inv(bigA)
    P_tt = (invbigA * Q.reshape(r2,1)).reshape(r,r);
    #=====
    for t in range(T):
        Xhat_t1t = matA*Xhat_tt;
        P_t1t    = matA*P_tt*matA.transpose() + Q;
        gamma_t  = matC.transpose()*P_t1t*matC;
        Yhat_t1t = matC.transpose()*Xhat_t1t;
        K_t1     = P_t1t*matC/gamma_t;
        I_t      = Y[t] - Yhat_t1t;
        Xhat_tt  = Xhat_t1t + K_t1*I_t;
        P_tt     = P_t1t - K_t1*matC.transpose()*P_t1t;
        s_t      = I_t**2/gamma_t;
        #===== current likelihood
```

```
        Lcurr    = Lcurr + log(gamma_t)+s_t;
    L = Lcurr/2;
    return L
#======= main program
# direct calculation and Kalman approach give
# identical values for the log-likelihood
N = 12;
#===== ARMA causal and inversible
a = array([1,-0.9,0.8]); b = array([1,0.6,0.1,0.3]);
sigma = 2; sigma2 = sigma**2;
w = sigma*randn(N); x = lfilter(b,a,w); matx = mat(x)
#===== with Kalman
Y = x; L1 = LLarmaKalman(Y,a,b,sigma2);
#===== direct ACFs
gamma,Ra = arma2ACF(a,b,sigma2,N);
Rgamma = toeplitz(gamma); invR = inv(Rgamma);
kappa = (matx*invR*matx.transpose());
L2 = (log(det(Rgamma))+kappa)/2;
print ('*******'); print(L1[0,0] / L2[0,0])
```

5.3.5.– (Filtering and smoothing for 2D tracking) (see p. 128)

1) Using a second-order Taylor expansion, for each component i, we have:

$$x_i(nT+T) \approx x_i(nT) + T\dot{x}_i(nT) + \frac{T^2}{2}\ddot{x}_i(nT) \qquad [5.35]$$

$$\dot{x}_i(nT+T) \approx \dot{x}_i(nT) + T\ddot{x}_i(nT) \qquad [5.36]$$

Hence:

$$X_{n+1} \approx \begin{bmatrix} 1 & 0 & T & 0 \\ 0 & 1 & 0 & T \\ 0 & 0 & 1 & 0 \\ 0 & 0 & 0 & 1 \end{bmatrix} X_n + \begin{bmatrix} T^2/2 & 0 \\ 0 & T^2/2 \\ T & 0 \\ 0 & T \end{bmatrix} \begin{bmatrix} \ddot{x}_1(nT) \\ \ddot{x}_2(nT) \end{bmatrix}$$

Considering that the pairs $(\ddot{x}_1(nT), \ddot{x}_2(nT))$ representing the acceleration form a series of independent, centered, Gaussian random variables with a covariance matrix $\sigma^2 I_2$, we have:

$$X_{n+1} = AX_n + B_n$$

with

$$R^B = \sigma^2 \begin{bmatrix} T^4/4 & 0 & T^3/2 & 0 \\ 0 & T^4/4 & 0 & T^3/2 \\ T^3/2 & 0 & T^2 & 0 \\ 0 & T^3/2 & 0 & T^2 \end{bmatrix}$$

σ is expressed in m/s^2.

2) Let the speed v_0 be 30 m/s, and let us consider that it may vary by a quantity proportional to v_0 of the form λv_0. We can therefore take $\sigma \approx \lambda v_0 / T$ as the dispersion of the acceleration.

3) Type and run the following program:

```
# -*- coding: utf-8 -*-
"""
Created on Sat Jun 11 17:50:17 2016
****** Kalmantraj2D
@author: maurice
"""
from numpy import zeros, size, cos, sin, arange
from numpy import pi, log, sqrt, eye, array, mod
from numpy import matrix as mat
from numpy.linalg import det, inv
from numpy.random import randn
from scipy.linalg import sqrtm
from matplotlib import pyplot as plt
#===============================================
def kalmanfilter(Y,A,RV,C,RW,mu0,R0):
    """
    # Kalman filter
    # SYNOPSIS
    #    [Xtt,Ptt,loglikeli] = kalmanfilter(Y,A,RV,C,RW,mu0,R0)
    # Inputs:
    #     Y   = observations (dimY x T)
    #     A   = state matrix (dimX x dimX)
    #     RV  = state covariance (dimX x dimX)
    #     RW  = observation covariance (dimY x dimY)
    #     mu0 = initial state mean (dimX x 1)
    #     R0  = initial state covariance (dimX x dimX)
    # Outputs:
    #     Xtt = filtered state (dimX x T)
    #     Ptt = covariance of estimate (dimX x dimX x T)
```

```
    #       loglikeli = log p(Y_1:T)
    """
    dimX            = size(A,0);
    T               = size(Y,1);
    Xtt             = mat(zeros([dimX,T]));
    Ptt             = zeros([dimX,dimX,T]);
    Ptt[:,:,0]      = 0;
    Pttm1           = zeros([dimX,dimX,T]);
    Xttm1           = A * mu0;
    Rttm1           = ((A * R0) * A.transpose()) + RV;
    cov_inov        = (C * Rttm1 * C.transpose()) + RW;
    invcov_inov     = inv(cov_inov)
    Kn              = (Rttm1*C.transpose())*invcov_inov;
    inov_1          = Y[:,0]-C*Xttm1;
    Xtt[:,0]        = Xttm1+Kn*inov_1;
    Pttm1[:,:,0]    = Rttm1;
    Ptt[:,:,0]      = Rttm1 - ((Kn*C)*Rttm1);
    loglikeli_k     = array(zeros(T));
    ll_k = -(log(det(cov_inov))
        + (inov_1.transpose()*invcov_inov)*inov_1)/2.0
    loglikeli_k[0] = ll_k[0,0].real
    for k in range(1,T):
        Xttm1           = A*Xtt[:,k-1];
        Rttm1           = (A*Ptt[:,:,k-1])*A.transpose() + RV;
        cov_inov        = (C*Rttm1)*C.transpose() + RW;
        invcov_inov     = inv(cov_inov)
        inov_k          = (Y[:,k] - (C*Xttm1));
        Kn              = (Rttm1*C.transpose())*invcov_inov;
        Xtt[:,k]        = Xttm1 + Kn*inov_k;
        Pttm1[:,:,k]    = Rttm1;
        Ptt[:,:,k]      = Rttm1 - (Kn*C)*Rttm1;
        ll_k = -(log(det(cov_inov))
            + (inov_1.transpose()*invcov_inov)*inov_1)/2.0
        loglikeli_k[k] = ll_k[0,0];
    loglikeli = sum(loglikeli_k);
    return Xtt, Ptt, loglikeli, Pttm1
#===========================================================
def rts(Y, A, RV, B, RW, mu0, R0):
    """
    # Kalman's smoother
    # HMM:
    #   X_{k+1} = A X_k+V_k, (evolution)
    #   Y_{k}   = B X_k+W_k, (observation)
```

```
    #
    #  parameters
    #      A, RV, B, RW, mu0, R0
    #========
    # Inputs:
    #    Y : observations
    #    A : state matrix
    #    RV : covariance of X
    #    B : observation matrix
    #    RW : covariance of Y
    #    mu0 : initial state mean
    #    R0 : initial state covariance
    # Outputs:
    #    XtT : smooth mean estimate
    #    PtT : smooth covariance estimate
    #    Xtt : filter mean estimate
    #    Ptt : filter covariance estimate
    """
    dimX = size(A,0);
    T = size(Y,1);
    XtT = mat(zeros([dimX,T]));
    PtT = zeros([dimX,dimX,T]);
    # call Kalman filter
    Xtt, Ptt, loglikeli, Pttm1 = \
        kalmanfilter(Y, A, RV, B, RW, mu0, R0);
    # Backward smoothing
    # initialize at time T
    XtT[:,T-1] = Xtt[:,T-1];
    PtT[:,:,T-1] = Ptt[:,:,T-1];
    for k in range(T-2,-1,-1):
        Xtt_k      = (Xtt[:,k]);
        Ptt_k      = (Ptt[:,:,k]);
        Pttm1_kp1 = (Pttm1[:,:,k+1]);
        # RTS recursion
        Gn = (Ptt_k * A.transpose()) * inv(Pttm1_kp1);
        XtT[:,k] = Xtt_k + Gn * (XtT[:,k+1]-(A*Xtt_k));
        PtT[:,:,k] = Ptt_k + Gn * (PtT[:,:,k+1] * Gn.transpose()
            - A * Ptt_k);
    return XtT, PtT, Xtt, Ptt, loglikeli
#================================================
def confidenceellipse(X0,Ce,alpha, col = 'g'):
    """
    # SYNOPSIS: CONFIDENCEELLIPSE(X0, E, c)
```

```
    # Ellipse equation:
    #      (X-X0)'*(C^-1)*(X-X0)= c(alpha)
    #      X0 = coordinates of the ellipse's center (2x1)
    #      C  = positive (2x2) matrix
    #      alpha = confidence level in(0,1)
    """
    Ne=100; theta = 2*pi*arange(0,Ne+1)/Ne;
    calpha = -2*log(1-alpha);
    Y = sqrt(calpha)*mat([cos(theta),sin(theta)])
    X = sqrtm(Ce)*Y
    xval = array(X[0,:]+X0[0])
    yval = array(X[1,:]+X0[1])
    plt.plot(xval[0,:],yval[0,:],col);
#================================================
T = 0.01; v0 = 30;
A = mat([[1.0, 0, T, 0],[0, 1.0, 0, T],\
     [0, 0, 1.0, 0],[0, 0, 0, 1.0]]);
C = mat([[1.0, 0.0, 0.0, 0.0],[0.0, 1.0, 0.0, 0.0]]);
D = mat([[T**2/2, 0],[0, T**2/2],[T, 0],[0, T]]);
#== trajectory generation
M0 = 2; N = 50; th = 2*pi*arange(0,N)/N/4.0;
Xtrue = mat([cos(th),cos(th)**2+sin(th)]); sigmaUdB = 25.0;
sigma2obs = 10**(-sigmaUdB/10.0);
#==== observation generation
RU = ((v0*T)**2)*sigma2obs*eye(2); Y = Xtrue + sqrtm(RU)*mat
(randn(2,N));
#==== a priori knowledge
mu = 0.005; sigma = mu*v0/T; RB = sigma*sigma*(D*D.transpose());
#==== initial conditions
mu0 = mat([Y[0,0],Y[1,0],0,0]).transpose(); R0=sigma*sigma*eye(4);
FilterKFresults = kalmanfilter(Y, A, RB, C, RU, mu0, R0);
Xfilt = FilterKFresults[0]; Pnn = FilterKFresults[1]; LL=FilterK
Fresults[2]
SmoothKFresults = rts(Y, A, RB, C, RU, mu0, R0);
Xsmooth = SmoothKFresults[0]; Pnnsmooth = SmoothKFresults[1];
LLsmooth = SmoothKFresults[2]
#====
Ya = array(Y); Xtruea = array(Xtrue); Xfilta = array(Xfilt);
Xsmootha = array(Xsmooth); alpha = 0.90
plt.clf(); plt.subplot(211)
plt.plot(Ya[0,:],Ya[1,:],'.-',color=[0.9,0.9,0.9]);
plt.hold('on')
plt.plot(Xtruea[0,:],Xtruea[1,:],'.-g')
```

```
plt.plot(Xtruea[0,0],Xtruea[1,0],'og',markersize=10)
plt.plot(Xfilta[0,:],Xfilta[1,:],'.-r');
plt.plot(Xsmootha[0,:],Xsmootha[1,:],'.-b'); plt.hold('off')
plt.subplot(212); plt.plot(Xfilta[0,:],Xfilta[1,:],'.-r');
plt.hold('on')
#plt.plot(Xtruea[0,0],Xtruea[1,0],'og',markersize=10)
plt.plot(Xsmootha[0,:],Xsmootha[1,:],'.-b'); plt.hold('off')
# we plot only a few of the confidence regions
for t in range(N):
   if mod(t,5)==1:
      plt.subplot(212); plt.plot(Xtruea[0,:],Xtruea[1,:],'.-g')
      plt.hold('on'); X0 = array([Xfilta[0,t],Xfilta[1,t]])
      Ce = mat(Pnn[0:2,0:2,t]); confidenceellipse(X0,Ce,alpha,col
         ='r');
      X0 = array([Xsmootha[0,t],Xsmootha[1,t]]);
      Ce = mat(Pnnsmooth[0:2,0:2,t])
      confidenceellipse(X0,Ce,alpha,col ='b');
      plt.xlim([0.8, 1.1]); plt.grid('on')
      plt.legend(('Kalman filter','true trajectory' ,'Kalman
         smoother'))
plt.hold('off'); plt.show()
```

REMARKS: modifying the value of σ using λ, we modify our initial ideas concerning the acceleration, and thus concerning the fact that the trajectory more or less follows a straight line. Hence, if σ is small, the mobile element is easy to follow when the trajectory is close to a straight line, but harder to follow if the trajectory curves. If σ is large, the observation noise is difficult to remove. The Kalman filter establishes a good compromise between prior knowledge, i.e. σ, and observations, i.e. Y_n.

5.3.6.– (Discrete HMM generation) (see p. 129)

1) Let us show that:

$$X_{n+1} = \sum_{j=0}^{S-1} j \times \mathbb{1}(U_n \in [F_{X_n}(j-1), F_{X_n}(j)])$$

where U_n is a series of independent r.v.s with values in $(0,1)$. To do this, let us determine $\mathbb{P}\{X_{n+1} = s | X_n = k\}$. As, according to our hypothesis, U_n is independent of X_n, we can write:

$$\mathbb{P}\{X_{n+1} = s | X_n = k\} = \mathbb{P}\{U_n \in [F_k(s-1), F_k(s)[\} = p_{k|s}$$

using the fact that U_n is uniform. Note that this expression may be written as:

$$X_{n+1} = f(X_n, U_n)$$

where U_n is an i.i.d. series with uniform distribution over $(0, 1)$. It is similar to the expression of the state equation evolution expression [3.9], except that it is neither linear nor Gaussian.

2) Type and run the program:

```
# -*- coding: utf-8 -*-
"""
Created on Thu Jun 30 05:56:00 2016
****** testHMMGgenerate
@author: maurice
"""
from numpy import zeros, array, size, cumsum, arange, pi, sqrt
from numpy import cos, sin, log
from scipy.linalg import sqrtm
from numpy.random import randn, rand
from numpy import matrix as mat
from matplotlib import pyplot as plt
#=================================================
def HMMGaussiangenerate(N,omega,P,mu,C):
    """
    # Generate an HMM of S gaussians
    # SYNOPSIS
    #     HMMGaussiangenerate(N,omega,P,mus,sigma2s)
    # inputs:
    #     N         = length of the sequence
    #     omega     = initial distribution S x 1
    #     P         = transition distribution S x S
    #     mu        = mean array d x S
    #     C         = variance array d x d x S
    # outputs:
    #     X         = state sequence 1 x N
    #     Y         = observation array d x N
    """
    d = size(C,0); S = size(P,0);
    C1on2 = zeros([d,d,S]);
    F = zeros([S,S]);
    for si in range(S):
        F[si,:]=cumsum(P[si,:]);
        C1on2[:,:,si] = sqrtm(C[:,:,si]);
```

```
    Fomega = cumsum(omega);
    Y = mat(zeros([d,N]));
    X = zeros(N);
    X[0] = S-int(sum(Fomega.flat>=rand()));
    Y[:,0] = mu[:,int(X[0])]+C1on2[:,:,int(X[0])]*mat(randn(d,1));
    for n in range(1,N):
        c=F[int(X[n-1]),:];
        X[n]= S-int(sum(c.flat>=rand()));
        Y[:,n]=mu[:,int(X[n])]+C1on2[:,:,int(X[n])]*mat(randn(d,1));
    return X,Y
def confidenceellipse(X0,Ce,alpha, col = 'g'):
    """
    # SYNOPSIS: CONFIDENCEELLIPSE(X0, E, c)
    # Ellipse equation:
    #    (X-X0)'*(C^-1)*(X-X0)= c(alpha)
    #    X0 = coordinates of the ellipse's center (2x1)
    #    C  = positive (2x2) matrix
    #    alpha = confidence level in(0,1)
    """
    Ne=100; theta = 2*pi*arange(0,Ne+1.0)/Ne;
    calpha = -2*log(1-alpha);
    Y = sqrt(calpha)*mat([cos(theta),sin(theta)])
    X = sqrtm(Ce)*Y
    xval = array(X[0,:]+X0[0])
    yval = array(X[1,:]+X0[1])
    plt.plot(xval[0,:],yval[0,:],col);
#== main
d = 2; S = 4; N = 3000;
Sigma2s = zeros([d,d,S]);
mus = mat(zeros([d,S]))
for ids in range(S):
    Maux = mat(randn(d,d))
    Sigma2s[:,:,ids] = Maux*Maux.transpose()
    mus[:,ids] = mat(randn(d,1));
P = mat(array([[0.4, 0.1, 0.3, 0.2], \
   [0.1, 0.4, 0.3, 0.2],[0.3, 0.1, 0.4, 0.2],[0.1, 0.3, 0.1, 0.5]]))
omega  =  mat(array([[1.0/2,1.0/4,1.0/8,1.0/8]]));
X,Y = HMMGaussiangenerate(N,omega,P,mus,Sigma2s);
plt.clf(); alpha = 0.95
for ids in range(S):
    Yids = array(Y[:,X==ids]); Xids = mus[:,ids]; Sids = Sigma2s
    [:,:,ids]
    plt.subplot(2,2,ids+1); plt.plot(Yids[0,:],Yids[1,:],'.')
```

```
        confidenceellipse(Xids,Sids,alpha, col = 'g')
plt.show()
```

5.3.7.– **(EM algorithm for HMM)** (see p. 136) Type the following module that includes the functions EMforHMM, HMMGaussiangenerate and ForwBackwGaussian:

```
# -*- coding: utf-8 -*-
"""
Created on Thu Jun 30 05:40:19 2016
****** toolEM
@author: maurice
"""
from numpy import zeros, size, exp,sum
from numpy import pi, log, sqrt, ones, cumsum
from numpy import matrix as mat
from numpy.linalg import det, pinv
from numpy.random import randn, rand
from scipy.linalg import sqrtm
def EMforHMM(Y,P_old,C_old,mu_old,alpha,beta,gamma,ct):
    """
    # SYNOPSIS:
    #    EMforHMM(Y, P_old,C_old,mu_old,alpha,beta,gamma,ct)
    # Inputs
    #    Y = array   d x N
    #    P_old = transition distribution S x S
    #        where P(j,i) = Prob(X_n = i|X_n-1 = j)
    #        independent of n
    #    C_old = observation covariance d x d x S
    #    mu_old = observation mean d x S
    #    alpha  = array N x 1
    #    beta   = array N x 1
    #    gamma  = array N x 1
    # Outputs
    #    omega_new,P_new,C_new,mu_new, ct
    """
    log2pi = log(2*pi);
    S = size(P_old,1);
    d = size(Y,0)
    N = size(Y,1)
    omega_new = gamma[:,0];
    xi = zeros([S,S,N-1]);
    g  = zeros([N,S]);
    for si in range(S):
```

```
            for nn in range(N):
                invC = pinv(C_old[:,:,si]);
                auxV = (Y[:,nn]-mu_old[:,si])
                aux_g = d*log2pi+log(det(C_old[:,:,si]))+ \
                    auxV.transpose()* invC * auxV
                g[nn,si] = exp(-aux_g/2);
        for nn in range(N-1):
            for sf in range(S):
                for si in range(S):
                    xi[sf,si,nn] = alpha[si,nn] * \
                    beta[sf,nn+1]* P_old[si,sf]*g[nn+1,sf];
            xi[:,:,nn] = xi[:,:,nn] / ct[nn+1];
        #=====
        sumgamma = sum(gamma,1);
        Paux = zeros([S,S]); P_new = zeros([S,S]);
        for si in range(S):
            for sf in range(S):
                Paux[si,sf]=sum(xi[sf,si,:]);
            P_new[si,:]=Paux[si,:]/sum(Paux[si,:]);
        #=====
        mu_new = mat(zeros([d,S])); C_new = zeros([d,d,S]);
        for si in range(S):
            mu_new[:,si]=Y*gamma[si,:].transpose()/sumgamma[si];
            C_newaux=0;
            for nn in range(N):
                auxV = Y[:,nn]-mu_new[:,si]
                C_newaux = C_newaux + (auxV*auxV.transpose())*gamma
                    [si,nn];
            C_new[:,:,si]=C_newaux/sumgamma[si];
        return omega_new,P_new,C_new,mu_new
#==============================================
def HMMGaussiangenerate(N,omega,P,mu,C):
    """
    # Generate an HMM of S gaussians
    # SYNOPSIS
    #    HMMGaussiangenerate(N,omega,P,mus,sigma2s)
    # inputs:
    #    N       = length of the sequence
    #    omega   = initial distribution S x 1
    #    P       = transition distribution S x S
    #    mus     = mean array d x S
    #    sigma2s = variance array d x d x S
    # outputs:
```

```
    #      X         = state sequence 1 x N
    #      Y         = observation array d x N
    """
    d = size(C,0); S = size(P,0);
    C1on2 = zeros([d,d,S]);
    F = zeros([S,S]);
    for si in range(S):
        F[si,:]=cumsum(P[si,:]);
        C1on2[:,:,si] = sqrtm(C[:,:,si]);
    Fomega = cumsum(omega);
    Y = mat(zeros([d,N]));
    X = zeros(N);
    X[0] = S-int(sum(Fomega.flat>=rand()));
    Y[:,0] = mu[:,int(X[0])]+C1on2[:,:,int(X[0])]*mat(randn(d,1));
    for n in range(1,N):
        c=F[int(X[n-1]),:];
        X[n]= S-int(sum(c.flat>=rand()));
        Y[:,n]=mu[:,int(X[n])]+C1on2[:,:,int(X[n])]*mat(randn(d,1));
    return X,Y
#========================================================
def ForwBackwGaussian(Y,omega,P,C,mu):
    """
    # SYNOPSIS:
    #          ForwBackwGaussian(Y,omega,P,C,mu)
    # Inputs:
    #     Y = array  d x N
    #     omega = intitial distribution array S x 1
    #     P     = transition distribution S x S
    #            where P(j,i) = Prob(X_n = i|X_n-1 = j)
    #            independent of n
    #     C     = observation covariance d x d x S
    #     mu    = observation mean d x S
    # Outputs:
    #     alpha = array N x 1 - forward propagation
    #     beta = array N x 1 - backward propagation
    #     gamma = array N x 1
    #     ell = log likelihood
    #     ct = normalisation constant N x 1
    """
    d = size(Y,0)
    N = size(Y,1)
    unsur2pid = (2*pi)**(-d/2);
    S = len(omega.flat);
```

```
        ct = zeros(N); alpha = zeros([S,N]);
        tildealpha = zeros(S); like = zeros([S,N]);
        for si in range(S):
            Cis = C[:,:,si]
            detCis = sqrt(det(Cis))
            muis = mu[:,si];
            for nn in range(N):
                yin  = Y[:,nn]-muis;
                aux1 = yin.transpose()*pinv(Cis)*yin;
                like[si,nn] = unsur2pid * exp(-aux1/2.0) / detCis;
        for si in range(S):
            tildealpha[si] = like[si,0]*omega.flat[si];
        ct[0] = sum(tildealpha);
        alpha[:,0] = tildealpha / ct[0];
        ell = log(ct[0]);
        for nn in range(1,N):
            gin = like[:,nn];
            for si in range(S):
                tildealpha[si] = gin[si] * alpha[:,nn-1].transpose()*
                P[:,si];
            ct[nn] = sum(tildealpha);
            alpha[:,nn] = tildealpha / ct[nn];
            ell = ell+log(ct[nn]);
        beta = ones([S,N]);
        for nn in range(N-2,0,-1):
            ginp1 = like[:,nn+1];
            for si in range(S):
                beta[si,nn]=sum((beta[:,nn+1] * P[si,:].flat) * ginp1)/
                ct[nn+1];
        gamma = mat(beta * alpha);
        return alpha,beta,gamma,ell,ct
```

Type and run the following program:

```
# -*- coding: utf-8 -*-
"""
Created on Sun Jun 12 06:52:34 2016
****** testEMforHMM
@author: maurice
"""
from numpy import zeros, eye, ones
from numpy import matrix as mat
from numpy.random import randn
from matplotlib import pyplot as plt
```

```
import toolEMforHMM as tEM
#===== testFB.m
# Uses function geneGMM.m to generate data
d=2; S=4; N=100; C = zeros([d,d,S]);
for si in range(S):
    C[:,:,si] = mat(eye(d));
mu = mat(randn(d,S));
P = mat([[0.4, 0.1, 0.3, 0.2],\
 [0.1, 0.4, 0.3, 0.2],[0.3, 0.1, 0.4, 0.2],[0.1, 0.3, 0.1, 0.5]]);
omega = mat([1.0/2,1.0/4,1.0/8,1.0/8]);
#===== testEMforHMMG.m
X,Y = tEM.HMMGaussiangenerate(N,omega,P,mu,C); mu_old = mat(randn(d,S));
#===== EM initialisation
omega_old = mat(ones(S)/S); C_old = zeros([d,d,S]);
for si in range(S):
    C_old[:,:,si] = mat(eye(d));
P_old = mat(ones([S,S])/S);
#=====
MAXITER = 200; LL=zeros(MAXITER); update_omega = 0;
ip = 0; THRESHOLD = 1e-8; llflag = 1; LL_previous = -1e7
while llflag:
    alpha, beta, gamma, ell, ct = \
        tEM.ForwBackwGaussian(Y,omega_old,P_old,C_old,mu_old);
    LL[ip] = ell; relgap = abs((LL[ip]-LL_previous)/LL_previous);
    LL_previous = LL[ip]
    #===
    [omega_new, P_new, C_new, mu_new] = \
        tEM.EMforHMM(Y,P_old,C_old,mu_old,alpha,beta,gamma,ct);
    omega_old=omega_new; P_old=mat(P_new); mu_old = mu_new; C_old
    = C_new;
    ip = ip+1; llflag = (relgap>THRESHOLD) & (ip<MAXITER)
LL = LL[range(ip)]; plt.clf(); plt.plot(LL,'.-'); plt.show()
```

5.3.8.– **(State estimation by Viterbi algorithm)** (see p. 139)

The following program test the Viterbi algorithm:

```
# -*- coding: utf-8 -*-
"""
Created on Sun Jun 12 11:24:24 2016
****** testViterby
@author: maurice
"""
```

```python
from numpy import zeros, size, argmax, log, array, cumsum
from numpy.random import randn, rand
#==================================================
def HMMGaussian1dgenerate(N,omega,P,mus,sigmas):
    """
    # Generate an HMM of S gaussians
    # SYNOPSIS
    #     HMMGaussiangenerate(N,omega,P,mus,sigma2s)
    # inputs:
    #     N        = length of the sequence
    #     omega    = initial distribution S x 1
    #     P        = transition distribution S x S
    #     mus      = mean array S x 1
    #     sigmas   = variance array S x 1
    # outputs:
    #     X        = state sequence N x 1
    #     Y        = observation array N x 1
    #     logG     = log-likelihood array N x S
    """
    S = size(P,0);
    F = zeros([S,S]);
    for si in range(S):
        F[si,:] = cumsum(P[si,:]);
    Fomega = cumsum(omega);
    Y = zeros(N); X = zeros(N); logG = zeros([N,S]);
    X[0] = S-int(sum(Fomega.flat>=rand()));
    sigmas_n = sigmas[int(X[0])]
    mus_n = mus[int(X[0])]
    Y[0] = mus_n+sigmas_n*randn();
    for n in range(1,N):
        c = F[int(X[n-1]),:];
        X[n] = S-int(sum(c.flat>=rand()));
        sigmas_n = sigmas[int(X[n])]
        mus_n = mus[int(X[n])]
        Y[n] = mus_n+sigmas_n*randn();
    for n in range(N):
        for si in range(S):
            mu_s = mus[si]; sigma_s = sigmas[si];
            logG[n,si]=-log(sigma_s)-(Y[n]-mu_s)**2/sigma_s/sigma_s
    return X, Y, logG
#======================================
def fviterbi(logG,logA):
    """
```

```
# Viterbi algorithm
# SYNOPSIS: hatX = FVITERBI(logG,logA)
# Inputs
#      logA(i,j) = log p(i|j)
#      logG(n,i) = array (N x S) = [ log g(y_n|i) ]
# Outputs
#      hatX = "optimal sequence"
"""
N = size(logG,0); S = size(logG,1);
met = zeros([N+1,S]); asc = zeros([N,S]);
hatX = zeros(N); d = zeros([S,S]);
for nn in range(N):
    for sj in range(S):
        for si in range(S):
            d[si,sj]= met[nn,sj]+logG[nn,si]+logA[si,sj];
    for si in range(S):
        met[nn+1,si] = max(d[si,:]);
        asc[nn,si] = argmax(d[si,:]);
hatX[N-1] = argmax(met[N-1,:]);
for nn in range(N-2,0,-1):
    hatX[nn] = asc[nn+1,int(hatX[nn+1])];
return hatX
#========== main program
S = 5; N = 12; Sigma2s = zeros(S); mus = zeros(S);
for ids in range(S):
    Sigma2s[ids] = 0.1*rand()
    mus[ids] = 10.0*randn();
P = array([[0.4, 0.1, 0.3, 0.2], \
    [0.1, 0.4, 0.3, 0.2],\
    [0.3, 0.1, 0.4, 0.2],\
    [0.1, 0.3, 0.1, 0.5]])
omega   = array([1.0/2,1.0/4,1.0/8,1.0/8]);
X, Y, logG = HMMGaussian1dgenerate(N,omega,P,mus,Sigma2s);
hatX = fviterbi(logG,P); print(sum(hatX == X))
```

5.4. Monte-Carlo methods

5.4.1.– **(Multinomial law)** (see p. 146) Type the following program:

```
# -*- coding: utf-8 -*-
"""
Created on Tue May 31 07:23:00 2016
****** multinomial
```

```
@author: maurice
"""
from numpy import zeros, cumsum, where
from numpy.random import rand
from matplotlib import pyplot as plt
N=10000; mu=[0.1,0.2,0.3,0.2,0.2]; lmu=len(mu); F=cumsum(mu);
x=zeros(N);
for ii in range(N):
    u = rand(); x[ii]=where(F>=u)[0][0]
plt.clf(); plt.hist(x, range(lmu+1),normed='True',rwidth=0.5);
plt.show()
```

5.4.2.– **(Homogeneous Markov chain)** (see p. 146) Type the following program:

```
# -*- coding: utf-8 -*-
"""
Created on Mon May 30 21:06:26 2016
****** MC
@author: maurice
"""
from numpy import array, cumsum, size, zeros, where
from numpy.random import rand
from matplotlib import pyplot as plt
# intial law
pi1 = array([0.5, 0.2, 0.3]); F1 = cumsum(pi1);
# transition matrix
A = array([[0.3, 0.0, 0.7],[0.1, 0.4, 0.5],[0.4, 0.2, 0.4]]);
La = size(A,0); N = 10000; X = zeros(N);
X[0] = len(F1[F1<rand()]); F = zeros([La,size(A,1)]);
for ia in range(La):
    F[ia,:] = cumsum(A[ia,:]);
for n in range(1,N):
    c = F[int(X[n-1]),:]; X[n] = sum(rand()>c)
plt.clf()
for ia in range(La):
    vcurr = where(X[range(N-1)]==ia)[0];
    succ  = X[vcurr+1]; plt.subplot(La,1,ia+1)
    plt.hist(succ,bins=(0,1,2,3),normed='True',rwidth=0.5);
plt.show()
```

5.4.3.– **(Linear transformation of 2D Gaussian)** (see p. 147) Type:

```
# -*- coding: utf-8 -*-
"""
```

```
Created on Mon May 30 15:17:06 2016
****** gauss2d
@author: maurice
"""
from numpy import array, dot
from scipy.linalg import eigh, sqrtm
from numpy.random import randn
from matplotlib import pyplot as plt
N = 5000; R = array([[2, 0.95],[0.95, 0.5]]);
S = sqrtm(R); W = randn(N,2); X = dot(W,S);
plt.clf(); plt.plot(X[:,0],X[:,1],'.')
plt.axis('square'); plt.xlim([-8, 8]); plt.ylim([-8, 8]),
[D,U] = eigh(R); plt.hold('on'),
plt.plot(U[1,1]*array([-8.0, 8.0]),-U[0,1]*array([-8.0, 8.0]),'r')
plt.plot(U[1,0]*array([-8.0, 8.0]),-U[0,0]*array([-8.0, 8.0]),'y')
plt.hold('off'); plt.show()
```

5.4.4.– (Box-Muller method) (see p. 148) The generation algorithm consists of the two following steps:

1) Draw two independent samples (U, V) with a uniform distribution over $(0, 1)$;

2) Calculate the variable pair:

$$\begin{cases} X = \sigma\sqrt{-2\log(U)}\cos(2\pi V) \\ Y = \sigma\sqrt{-2\log(U)}\sin(2\pi V) \end{cases}$$

Type the following program:

```
# -*- coding: utf-8 -*-
"""
Created on Mon May 30 05:36:27 2016
#****** boxmuller
@author: maurice
"""
from numpy.random import rand
from numpy import cos, sin, sqrt, log, pi
from matplotlib import pyplot as plt
from matplotlib.colors import LogNorm
N = 100000; sigma = 1; U = rand(N); V = rand(N);
X = sigma*sqrt(-2*log(U)) * cos(2*pi*V); Y = sigma*sqrt(-2*log(U)) * sin(2*pi*V);
plt.clf(); plt.hist2d(X,Y, bins=40, norm=LogNorm()); plt.show()
```

5.4.5.– (The Cauchy distribution) (see p. 148)

1) The cumulative function of Z is written as:

$$\mathbb{P}\{Z \leq z\} = \frac{1}{2} + \frac{1}{\pi}\operatorname{atan}\frac{z-z_0}{a}$$

The density is therefore expressed as:

$$p_Z(z) = \frac{a}{\pi(a^2 + (z-z_0)^2)}$$

2) The cumulative function of Z is written as:

$$\mathbb{P}\{Z \leq z\} = \int_{\{(x,y):ay/x<z-z_0\}} \frac{1}{2\pi}e^{-(x^2+y^2)/2}dxdy$$

$$\mathbb{P}\{Z \leq z\} = \frac{1}{2\pi}\int_{-\infty}^{0} e^{-x^2/2} \int_{x(z-z_0)/a}^{+\infty} e^{-y^2/2}dydx$$

$$+ \frac{1}{2\pi}\int_{0}^{+\infty} e^{-x^2/2} \int_{-\infty}^{x(z-z_0)/a} e^{-y^2/2}dydx$$

$$= \frac{1}{2\pi}\int_{0}^{+\infty} e^{-x^2/2} \int_{-x(z-z_0)/a}^{+x(z-z_0)/a} e^{-y^2/2}dydx$$

From the derivative with respect to z, we obtain the density:

$$p_Z(z) = \frac{1}{a\pi}\int_{0}^{+\infty} xe^{-x^2(1+(z-z_0)^2/a^2)/2}dx$$

$$= \frac{1}{a\pi}\int_{0}^{+\infty} e^{-u(1+(z-z_0)^2/a^2)}du = \frac{1}{\pi}\frac{a}{a^2+(z-z_0)^2}$$

3) Type and run the following program:

```
# -*- coding: utf-8 -*-
"""
Created on Mon May 30 05:56:42 2016
#****** cauchylaw
@author: maurice
"""
from numpy import tan, pi, linspace
from numpy.random import randn, rand
from matplotlib import pyplot as plt
from scipy.stats import probplot, cauchy
```

```
z0=10; a=0.5; N=1000; z=linspace(-5.0,25.0,50);
pztheo = cauchy.pdf(z,z0,a);
#===== atan(U)
z1 = z0+a*tan(pi*(rand(N)-0.5));
#===== Y/X
X = randn(N,2); z2 = z0+a*(X[:,1] / X[:,0]);
plt.clf(); plt.subplot(221); plt.hist(z1,z,normed='True');
plt.hold('on'); plt.plot(z,pztheo,'.-r'); plt.hold('off'); plt.
yticks([])
plt.subplot(222); probplot(z1, dist="cauchy", plot=plt)
plt.xticks([]); plt.yticks([]); plt.title('')
plt.subplot(223); plt.hist(z2,z,normed='True');
plt.hold('on'); plt.plot(z,pztheo,'.-r');
plt.hold('off'); plt.yticks([])
plt.subplot(224); probplot(z1, dist="cauchy", sparams=(z0,a),
plot=plt)
plt.xticks([]); plt.yticks([]); plt.title('');
plt.show()
```

5.4.6.– (Metropolis-Hastings algorithm) (see p. 154)

Type the following program:

```
# -*- coding: utf-8 -*-
"""
Created on Sun May 29 21:08:08 2016
****** applicationMetropolis
@author: maurice
"""
from numpy import zeros, mean
from numpy import random
from scipy.stats import norm
from matplotlib import pyplot as plt
N = 10000; sigma = 2; U0 = 10*sigma; x = zeros(N); pprevious = 1;
for n in range(1,N):
    xproposal = U0*(random.rand()-0.5);
    pproposal = norm.pdf(xproposal,0.0,sigma);
    rho = pproposal / pprevious;
    if rho > 1:
        x[n] = xproposal; pprevious = pproposal;
    else:
        b = random.rand()<rho;
        x[n] = xproposal*b+x[n-1]*(1.0-b);
        pprevious = pproposal*b+pprevious*(1.0-b);
```

```
Nburn = 200; Nval = N - Nburn; x = x[Nburn:N];
#===== integral approx. value
I = mean(x ** 2); plt.clf()
auxhist=plt.hist(x,bins=30,normed='True'); dx=auxhist[1]
gtheo = norm.pdf(dx,0.0,sigma)
plt.hold('on'); plt.plot(dx,gtheo,'o-r',linewidth=2)
plt.hold('off'); plt.title('Approx value of I=%4.5f'%I);plt.show()
```

5.4.7.– **(Gibbs sampler)** (see p. 155)

1) The conditional distribution is given by expression [1.46]:

$$p_{X_1|X_2}(x_1,x_2) = \mathcal{N}\left(\mu_1 + \rho\frac{\sigma_1}{\sigma_2}(X_2-\mu_2), \sigma_1^2(1-\rho^2)\right)$$

2) Type the following program:

```
# -*- coding: utf-8 -*-
"""
Created on Sun May 29 17:56:22 2016
#****** applicationGibbs
@author: maurice
"""
from numpy import array, zeros, sqrt, mean, dot, size
from matplotlib import pyplot as plt
from numpy import random
mus=[2,1]; sigmas=[2,3]; rho = 0.9; C12 = rho*sigmas[0]*sigmas[1];
C = array([[sigmas[0]**2, C12],[C12, sigmas[1]**2]])
N = 10000; X = zeros([N,2]);
for n in range(1,N):
    mucond = mus[0]+rho*sigmas[0]*(X[n-1,1]-mus[1])/
    sigmas[1];
    sigmacond = sigmas[0]*sqrt(1.0-rho*rho);
    X[n,0] = mucond+sigmacond*random.randn();
    mucond = mus[1]+rho*sigmas[1]*(X[n,0]-mus[0])/sigmas[0];
    sigmacond = sigmas[1]*sqrt(1.0-rho*rho);
    X[n,1] = mucond+sigmacond*random.randn();
Nburn = 200; Nval = N - Nburn;
X = X[Nburn:N,:]; Xc = zeros([size(X,0),size(X,1)])
Xc[:,0] = X[:,0] - mean(X[:,0]); Xc[:,1] = X[:,1] - mean(X[:,1]);
Cestim = dot(Xc.transpose(),Xc)/(Nval-1)
for i1 in range(2):
    for i2 in range(2):
        print('C[%i,%i] = %4.3e, estimate C[%i,%i] = %4.3e'%(i1,i2,
```

```
        C[i1,i2]\
            ,i1,i2,Cestim[i1,i2]))
plt.clf(); plt.plot(X[:,0],X[:,1],'.'); plt.show()
```

5.4.8.– **(Importance sampling)** (see p. 160)

Type the following program:

```
# -*- coding: utf-8 -*-
"""
Created on Mon May 30 17:21:28 2016
****** IS_GaussfromCauchy
@author: maurice
"""
#=====
# Importance sampling for estimation of
# int_{alpha}^{+infty} p(x)dx with p(x)=N(0,1)
#     (1) directly from N(0,1)
#     (2) from Cauchy distribution and weights
#     (3) from Cauchy without knowing p(x) up to
#         a multiplicative constant
#=====
from numpy import zeros, sqrt, pi, tan, sum, exp
from numpy import std, linspace
from numpy.random import randn, rand
from matplotlib import pyplot as plt
from scipy.stats import norm
N = 10000; alpha = 3.0; palpha_th = 1-norm.cdf(alpha,0.0,1.0);
palpha_th2=palpha_th*palpha_th; Lruns=500; P1_direct=zeros(Lruns);
P1_IS_withCteNorm1=zeros(Lruns);P1_IS_withoutCteNorm1=zeros(Lruns);
dpitdemi = sqrt(2/pi);
for ii in range (Lruns):
    x = randn(N,1); u = rand(N,1); x_IS = tan(pi*(u-0.5));
    P1_direct[ii] = sum(x>alpha)/float(N); x_IS2 = x_IS**2;
    weight_IS_withoutCteNorm1 = exp(-(x_IS2) /2) * (1.0+x_IS2);
    P1_IS_withoutCteNorm1[ii] = sum((x_IS>alpha) *
            (weight_IS_withoutCteNorm1/ \
            sum(weight_IS_withoutCteNorm1)));
    poids_IS_withCteNorm1 = \
        weight_IS_withoutCteNorm1/dpitdemi;
    P1_IS_withCteNorm1[ii] = \
        sum((x_IS>alpha)*poids_IS_withCteNorm1/float(N));
plt.clf(); plt.subplot(1,3,1)
plt.hist(P1_direct, normed='true');
```

```
plt.xticks(linspace(0,0.0025,3),fontsize=8)
plt.subplot(1,3,2); plt.hist(P1_IS_withCteNorm1, normed='true');
plt.xticks(linspace(0,0.0025,3),fontsize=8)
plt.subplot(1,3,3)
plt.hist(P1_IS_withoutCteNorm1, normed='true');
plt.xticks(linspace(0,0.0025,3),fontsize=8)
ect_direct_estime = std(P1_direct-palpha_th);
ect_direct_theo   = sqrt((palpha_th-palpha_th2)/N);
ect_IS_estime     = std(P1_IS_withCteNorm1-palpha_th);
ect_IS_theo       = sqrt((sqrt(pi)*3*(1-norm.cdf(alpha*sqrt(2.0)))/
4.0+\
    +0.25*alpha*exp(-alpha*alpha)-palpha_th2)/N); plt.show()
print('Draw from Gaussian distribution')
print('\tTheorical std = %5.2e, Estimated std = %5.2e' \
    %(ect_direct_theo,ect_direct_estime));
print('Draw from Cauchy distribution and IS (not normalized)')
print('\tTheorical std = %5.2e, Estimated std = %5.2e'\
    %(ect_IS_theo,ect_IS_estime));
```

The results show that the direct calculation is more dispersive.

5.4.9.– (**Stratification**) (see p. 164)

1) Following [1.35] for $\mu = 0$ and $\sigma^2 = 1$: $\mathbb{E}\{e^{juX}\} = e^{-u^2/2} = \mathbb{E}\{\cos(ux) + j\sin(ux)\}$. Thus, $\mathbb{E}\{\cos(uX)\} = e^{-u^2/2}$.

2) Type the program:

```
# -*- coding: utf-8 -*-
"""
Created on Tue May 31 20:51:35 2016
****** stratification
@author: maurice
"""
from numpy import exp, sum, cos, mean, ones, zeros, std
from numpy.random import rand
from scipy.stats import norm
from matplotlib import pyplot as plt
# choose a value of u
u=3.0; I_exact = exp(-u**2/2);
#===== stratif_uniform function
def stratif_uniform(vect_n):
# SYNOPSIS: STRATIF_UNIFORM(vect_n)
# vect_n = k-length sequence to be drawn uniformly
```

```
# in the intervals of length 1/k of [0,1]
    k = len(vect_n); ampl = 1.0/k;
    N = int(sum(vect_n)); U = zeros(N); Sjm1 = 0; Sj = -1;
    for j in range(k):
        vj = int(vect_n[j]); Sj = Sj+vj;
        U[Sjm1:Sj+1] = (rand(vj)+(j))*ampl; Sjm1 = Sj+1;
    return U
#======= main program
N = 1000; S = 200; ni = int(N/S); vect_n = ones(S)*ni;
nb_runs = 3000; hatDirect_I = zeros(nb_runs);
hatStratif_I = zeros(nb_runs);
for i_run in range(nb_runs):
    Udirect = rand(N); Xdirect = norm.isf(Udirect,0,1);
    hatDirect_I[i_run] = sum(cos(u*Xdirect))/N;
    Ustrate = stratif_uniform(vect_n); Xstrate = norm.isf
    (Ustrate,0,1);
    hatStratif_I[i_run] = sum(cos(u*Xstrate))/N;
plt.clf()
plt.boxplot([hatDirect_I, hatStratif_I])
plt.show()
td = 'bias DIRECT = %5.2e, std DIRECT = %5.2e'\
    %(abs(I_exact-mean(hatDirect_I)),std((hatDirect_I)));
ts = 'bias STRATIF = %5.2e, std STRATIF = %5.2e'\
    %(abs(I_exact-mean(hatStratif_I)),std((hatStratif_I)));
print('****************************')
print(td); print(ts)
```

5.4.10.– (Antithetic variate approach) (see p. 164)

1) Using the estimators given by [4.14] and [4.32], respectively, we have

$$\text{cov}(f(X),f(X)) = \int_0^1 \frac{1}{(1+x)^2}dx - I^2 = 0.5 - \log^2(2)$$

$$\text{cov}(f(X),f(X)) = \int_0^1 \frac{1}{(2-x)(2-x)}dx - I^2 = 0.5 - \log^2(2)$$

and

$$\text{cov}\left(f(X),f(\widetilde{X})\right) = \int_0^1 \frac{1}{(1+x)(2-x)}dx - I^2 = \frac{2}{3}\log(2) - \log^2(2)$$

For $N = 100$, we therefore obtain $\text{var}\left(\widehat{I}_1\right) \approx 1.95\,10^{-4}$ and $\text{var}\left(\widehat{I}_a\right) \approx 1.19\,10^{-5}$.

2) Type the program:

```
# -*- coding: utf-8 -*-
"""
Created on Sun May 29 17:27:37 2016
#****** antithetic
@author: maurice
"""
from numpy import log, std, mean
from numpy.random import rand
Lruns = 2000; N = 100; Ns2 = int(N/2); X = rand(N,Lruns);
Xtilde = 1.0-X; fX = 1.0 / (1.0+X); fXtilde = 1.0 / (1.0+Xtilde);
I1 = mean(fX,axis=0);
I2 = (mean(fX[1:Ns2,:],axis=0)+mean(fXtilde[1:Ns2,:],axis=0))/2.0;
Nvar1th = 0.5-log(2.0)*log(2.0); Nvar2th = 2.0*log(2.0)/3.0-log
(2.0)*log(2.0);
print('theoretical variances: %4.2e(direct), %4.2e(antith.)' \
    %(Nvar1th/N, (Nvar1th+Nvar2th)/N));
print('empirical variances: %4.2e(direct), %4.2e(antith.)' \
    %(std(I1)**2, std(I2)**2));
```

Bibliography

[BAS 93] BASSEVILLE M., NIKIFOROV I., *Detection of Abrupt Changes: Theory and Application*, Prentice-Hall, Upper Saddle River, 1993.

[BIL 12] BILLINGSLEY P., *Probability and Measure*, John Wiley & Sons, 2012.

[BLA 14] BLANCHET G., CHARBIT M., *Digital Signal and Image Processing Fundamentals*, vol. 1, 2nd ed., ISTE, London and John Wiley & Sons, New York, 2014.

[BRO 90] BROCKWELL P., DAVIES R., *Time Series: Theory and Methods*, Springer Verlag, 1990.

[BUR 02] BURNHAM K.P., ANDERSON D.R., *Model Selection and Multimodel Inference: A Practical Information-Theoretic Approach*, Springer, New York, 2002.

[CAP 05] CAPPÉ O., MOULINES E., RYDEN T., *Inference in Hidden Markov Models*, Springer New York, 2005.

[DEM 77] DEMPSTER A.P., LAIRD N.M., RUBIN D.B., "Maximum likelihood from incomplete data via the EM algorithm", *Journal of Royal Statistical Society: Series B*, vol. 39, pp. 1–38, 1977.

[EFR 79] EFRON B. "Bootstrap methods: another look at the jackknife", *Annals of Statistics*, vol. 7, no. 1, pp. 1–26, 1979.

[GEL 08] GELMAN A., "Scaling regression inputs by dividing by two standard deviations", *Statistics in Medecine*, vol. 27, pp. 2865–2873, 2008.

[HAS 09] HASTIE T.J., TIBSHIRANI R.J., FRIEDMAN J.H., *The Elements of Statistical Learning: Data Mining, Inference, and Prediction*, Springer, New York, 2009.

[EMI 60] EMIL KALMAN R., "A new approach to linear filtering and prediction problems", *Transactions of the ASME – Journal of Basic Engineering*, vol. 82, pp. 35–45, 1960.

[MAT 98] MATSUMOTO M., NISHIMURA T., "Mersenne twister: a 623-dimensionally equidistributed uniform pseudorandom number generator", *ACM Transactions on Modeling and Computer Simulations*, vol. 8, no. 1, pp. 3–30, 1998.

[MC 43] MCCLURE F.J., "Ingestion of fluoride and dental caries: quantitative relations based on food and water requirements of children one to twelve years old", *American Journal of Diseases of Children*, vol. 66, no. 4, pp. 362–369, 1943.

[MIL 74] MILLER R.G., "The jacknife – a review", *Biometrika*, vol. 61, no. 1, pp. 1–15, 1974.

[MON 10] MONTGOMERY D.C., RUNGER G.C., *Applied Statistics and Probability for Engineers*, John Wiley & Sons, 2010.

[NEY 33] NEYMAN J., PEARSON E.S., "On the problem of the most efficient tests of statistical hypotheses", *Philosophical Transactions of the Royal Society A: Mathematical, Physical and Engineering Sciences*, vol. 231, pp. 694–706, 1933.

[PEA 88] PEARL J., *Probabilistic Reasoning in Intelligent Systems: Networks of Plausible Inference*, Morgan Kaufmann, 1988.

[QUE 56] QUENOUILLE M.H., "Notes on bias in estimation", *Biometrika*, vol. 43, 1956.

[RUD 86] RUDIN W., *Real and Complex Analysis*, 3rd ed., McGraw-Hill, 1986.

[SHA 08] SHARMA A.K., *Textbook of Biostatistics: Volume 1*, Discovery Publishing, 2008.

[STI 73] STIGLER S., "Studies in the history of probability and statistics. XXXII: Laplace, Fisher and the discovery of the concept of sufficiency", *Biometrika*, vol. 60, no. 3, pp. 439–445, 1973.

[TUC 58] TUCKEY J.W., "Bias and confidence in not-quite large samples", *Ann. Math. Statist.*, vol. 29, pp. 614–623, 1958.

Index

A, B

acceptance-rejection method, 149
AIC, 76
algorithm
    acceptance-rejection, 149
    backward recursion, 133
    Box-Muller, 148
    EM for GMM, 94
    Expectation-Maximization (EM), 91
    forward recursion, 132
    Gibbs sampler, 155
    importance sampling, 158
    Kalman, 228
    LDA, 40
    Metropolis-Hastings, 153, 154
    PCA, 36
    Viterbi, 138
antithetic variates method, 164
AR, 125
area under the ROC curve (AUC), 49
AUC, 49
autoregressive, 125
backward stepwise selection, 80
Bayesian statistics, 59
best linear unbiased estimator (BLUE), 70
bias, 59
BIC, 76
bilateral hypothesis testing, 52
bilateral test, 44
bootstrap, 107
box plot, 30
Box-Muller method, 148
burn-in period, 151

C

Cauchy, 148
    generation, 148
    law, 148
    moment, 148
cdf, 2, 4
censored data, 97
centered
    random variable, 7
central limit
    theorem, 22
characteristic function, 6
    marginal probability distribution, 6
classification, 36, 41, 88, 196
coefficients
    correlation, 7
complete data (EM algorithm), 91
composite hypothesis, 43
conditional
    covariance (Gaussian case), 17
    distribution (Gaussian case), 17
    expectation, 10, 14
    expectation (Gaussian case), 17
    probability distribution, 10
confidence interval, 55
continuity
    theorem, 24
correlation, 7

coefficients, 7
distribution, 199
matrix, 8
covariance, 7
Cramer-Rao bound (CRB), 60, 93
critical
region, 44
test function, 44
cross validation, 107, 111
cumulated sum, 57
cumulative
distribution, 4
function, 2, 33, 55, 58, 98, 103, 104, 106, 144
CUSUM, 57

D, E, F

δ-method, 20
density (probability), 3
design matrix, 65, 112
detailed balance equation, 152
detection probability, 44
deterministic test, 46
digamma function, 201
directed acyclic graph, 114
efficient, 61, 77
equation
detailed balance, 152
evolution, 119
observation, 119
state, 119
estimator, 58
efficient, 60, 77, 85
unbiased, 60, 61, 70, 71, 77, 81, 85, 109
Expectation-Maximization, 91
false alarm probability, 44
filtering, 117
Fisher
information matrix, 60
transformation, 89, 199
"fresh" data, 73
function
characteristic, 6
digamma, 201
EMforHMM, 249
estimparamGMM.py, 207

fgmm, 192
fminsearch, 83
ForwBackwGaussian, 249
HMMGaussiangenerate, 249
LLarmaKalman, 234
pcaldatoolbox.py, 170
toolGMM.py, 204
toollogistic.py, 212

G, H, I

Gaussian mixture model (GMM), 93
generalized expectation maximization (GEM), 92
generalized likelihood ratio test (GLRT), 51, 52, 54, 55, 57, 89, 102
generalized method of moments, 82
generation
Cauchy, 148
cumulative function inversion, 144
Gauss, 148
Gauss 2D, 147
Markov chain, 146
multinomial, 146
Rayleigh, 145
Gibbs sampler, 151, 155
goodness of fit test, 57
hidden Markov model, 113
histogram, 30, 53, 58, 103
HMM generation, 130
homogeneous Markov chain, 146
hypothesis
composite, 43
simple, 43
test, 43
identifiable, 42
importance
sampling, 157
weights, 157
incomplete data (EM algorithm), 91
inequality
Schwarz, 11
innovation, 121, 227
instrumental distribution, 156
intercept, 65
inversion method, 144

J, K, L

jackknife, 107
Jacobian, 18, 19
   useful formula, 19
joint probability, 4
Kalman
   filter, 121, 226
   gain, 122
   innovation, 122
   prediction, 121
   update, 121
kernel (smoothing), 103
Lagrange multipliers, 36
law
   Cauchy, 148
   Gauss, 148
   Gaussian, 147
   large numbers, 30
   multinomial, 146
   Rayleigh, 145
   transition, 151
   uniform, 144
LDA, 34, 37
least squares
   assumption, 65
   design matrix, 65
   method, 62
   null space, 67
   ordinary, 81
   unbiased, 70, 71
   useful notations, 66
   weighted, 81
leave one out (LOO), 111
leverage, 66, 72
likelihood, 43
   ratio, 45
linear discriminant analysis, 34, 37
linear regression, 65
logistic, 100, 102
Lyapunov equation, 234

M, N, O

Mahalanobis distance, 88
Mann-Whitney statistics, 50
marginal probability distribution
   characteristic function, 6

Markov
   chain, 146, 151
   homogeneous, 151
   process, 151
   transition law, 151
matrix
   covariance, 8
maximum likelihood
   method, 84
   estimator, 84
MCMC, 144, 151
mean, 7
   efficient estimator, 61, 77
   vector, 8
measurement vector, 118
median, 2
   empirical, 32, 106
Mersenne twister, 144
method
   acceptance-rejection, 149
MLE, 84
   i.i.d. gaussian, 87
   linear model, 88, 194
model
   Gaussian mixture (GMM), 93
   hidden Markov, 113
   parametric, 41
   statistical, 41
model selection, 74, 76
moments
   method, 81
monolateral test, 44
Monte-Carlo
   Markov chain, 144
   methods, 141
noise
   measurement, 120
   model, 120
observability, 123
observation equation, 119
OLS, 81
order statistics, 217
ordinary least squares, 81
orthogonal projection, 13
orthogonality principle, 12
overtraining, 222

P

parameter estimation, 58
parametric model, 41
path metrics, 137
pivotal, 74
plot
    boxplot, 29
    hist, 29
    qqplot, 29
    scatterplot, 29
prediction, 117
    step, 117
predictors, 65
principal components, 147
    analysis (PCA), 34, 36
probability
    density, 3
    distribution, 1
    joint, 3
probability density, 3
    joint, 4
probability distribution
    Gaussian, 14
    normal, 14
    sum of two r.v., 19
program
    afewdistributions.py, 26
    antithetic.py, 260
    applicationGibbs.py, 256
    applicationMetropolis.py, 255
    bootstraponmean.py, 109
    bootstraponregression.py, 219
    boxmuller.py, 253
    BSSdiabetes.py, 224
    cauchylaw.py, 254
    censoredhearttransplantation.py, 209
    censoredHT.py, 209
    censoredHTwithexogenous.py, 211
    chi2test.py, 181
    co2CV.py, 222
    CO2linmod.py, 184
    correlationdistribution.py, 199
    cumulEstimate.py, 216
    CumulInverseEstimate.py, 217
    CUSUMrecursiveformula.py, 179
    CUSUMtest.py, 180
    deltaMethodRice.py, 168
    diabetesvalidationmodel.py, 189
    diabetesZscore.py, 188
    egalizeimage.py, 219
    estimdsproba.py, 104
    estimf0.py, 63
    estimMarkovchain.py, 203
    estimStateGMM.py, 208
    gauss2d.py, 252
    gaussianquantiles.py, 2
    generate2Dtrajectory.py, 129
    histograminbrief.py, 31
    ICpercent.py, 169
    irisclassification.py, 195
    irispcalda.py, 172
    IS_GaussfromCauchy.py, 257
    Kalmantraj2D.py, 238
    KFnoisyAR1.py, 229
    leverageeffect.py, 184
    logisticORing.py, 214
    logisticOwner.py, 215
    logisticOwnerCV.py, 223
    logistictestGLRT.py, 215
    MC.py, 252
    MLEexponential.py, 200
    MMmixture.py, 192
    MMversusMLEcorrelation.py, 197
    modlinchangeNILE.py, 185
    multinomial.py, 251
    orderEstimCV.py, 220
    qqplotexample.py, 33
    rayleighsimul.py, 145
    regagediabetes.py, 186
    regboston.py, 187
    roccurve2gaussians.py, 173
    scatterplotexample.py, 29
    stratification.py, 258
    studentlawdiffm0.py, 174
    studentlawdiffm0m1.py, 178
    symbolicforCRBGaussian.py, 183
    testcorrelationWH.py, 198
    testEMforHMMG, 249
    testfluorcaries.py, 78
    testHMMGgenerate.py, 243
    testKarmaOnARMApq.py, 234
    testViterbi.py, 249
    kalmantraj1D.py, 230

Ttest.py, 177
verifyCRBrho.py, 196
proposition distribution, 156
p-value, 54

Q, R, S

QQ-plot, 33
qqplot, 29, 32
quadratic risk, 59
quantile, 10, 33, 106
R^2, 33
    adjusted, 75
random
    vector, 7
random process
    AR, 125
random value
    mean, 7
    variance, 7
random variable
    Gaussian, 14
    independence, 4
    standard deviation, 7
randomized test, 45
ranking, 41
Rayleigh, 145
regression, 62, 100
    logistic, 102
regression analysis, 41, 62
Riccati equation, 122
right-censored data, 97
ROC curve, 48
scatterplot, 29
seasonal trend, 30
significance level, 44
simple hypothesis, 43

simpy, 61
smoothing, 117
standardization, 65, 76, 79
state, 118
    vector, 118
stationary (Markov chain), 152
statistic of test, 44
statistical inference, 40
statistical model, 41
statistics (estimate), 43
symbol
    proportional to, 118
symbolic calculus, 61

U, V, W, Z

UMP, 44
unbiased, 60, 61, 70, 71, 77, 81, 85, 109
uniform distribution, 144
uniformly most powerful, 44
unilateral hypothesis testing, 52
update step, 117
variable
    dependent, 41
    explained, 41
    explanatory, 41, 65
    independent, 41, 65
    response, 41, 65
Viterbi
    backtracking, 138
    lattice, 138
Voronoi regions, 88
weighted least squares, 81
white
    random sequence, 10
WLS, 81
Z-score, 74

Other titles from

in

Digital Signal and Image Processing

2015

BLANCHET Gérard, CHARBIT Maurice
Digital Signal and Image Processing using MATLAB®
Volume 2 – Advances and Applications: The Deterministic Case – 2nd edition
Volume 3 – Advances and Applications: The Stochastic Case – 2nd edition

CLARYSSE Patrick, FRIBOULET Denis
Multi-modality Cardiac Imaging

DUBUISSON Séverine
Tracking with Particle Filter for High-dimensional Observation and State Spaces

GIOVANNELLI Jean-François, IDIER Jérôme
Regularization and Bayesian Methods for Inverse Problems in Signal and Image Processing

MAÎTRE Henri
From Photon to Pixel: The Digital Camera Handbook

2014

AUGER François
Signal Processing with Free Software: Practical Experiments

BLANCHET Gérard, CHARBIT Maurice
Digital Signal and Image Processing using MATLAB®
Volume 1 – Fundamentals – 2nd edition

DUBUISSON Séverine
Tracking with Particle Filter for High-dimensional observation and State Spaces

ELL Todd A., LE BIHAN Nicolas, SANGWINE Stephen J.
Quaternion Fourier Transforms for Signal and Image Processing

FANET Hervé
Medical Imaging Based on Magnetic Fields and Ultrasounds

MOUKADEM Ali, OULD Abdeslam Djaffar, DIETERLEN Alain
Time-Frequency Domain for Segmentation and Classification of Non-stationary Signals: The Stockwell Transform Applied on Bio-signals and Electric Signals

NDAGIJIMANA Fabien
Signal Integrity: From High Speed to Radiofrequency Applications

PINOLI Jean-Charles
Mathematical Foundations of Image Processing and Analysis Volumes 1 and 2

TUPIN Florence, INGLADA Jordi, NICOLAS Jean-Marie
Remote Sensing Imagery

VLADEANU Calin, EL ASSAD Safwan
Nonlinear Digital Encoders for Data Communications

2013

GOVAERT Gérard, NADIF Mohamed
Co-Clustering

DAROLLES Serge, DUVAUT Patrick, JAY Emmanuelle
Multi-factor Models and Signal Processing Techniques: Application to Quantitative Finance

LUCAS Laurent, LOSCOS Céline, REMION Yannick
3D Video: From Capture to Diffusion

MOREAU Eric, ADALI Tulay
Blind Identification and Separation of Complex-valued Signals

PERRIN Vincent
MRI Techniques

WAGNER Kevin, DOROSLOVACKI Milos
Proportionate-type Normalized Least Mean Square Algorithms

FERNANDEZ Christine, MACAIRE Ludovic, ROBERT-INACIO Frédérique
Digital Color Imaging

FERNANDEZ Christine, MACAIRE Ludovic, ROBERT-INACIO Frédérique
Digital Color: Acquisition, Perception, Coding and Rendering

NAIT-ALI Amine, FOURNIER Régis
Signal and Image Processing for Biometrics

OUAHABI Abdeljalil
Signal and Image Multiresolution Analysis

2011

CASTANIÉ Francis
Digital Spectral Analysis: Parametric, Non-parametric and Advanced Methods

DESCOMBES Xavier
Stochastic Geometry for Image Analysis

FANET Hervé
Photon-based Medical Imagery

MOREAU Nicolas
Tools for Signal Compression

2010

NAJMAN Laurent, TALBOT Hugues
Mathematical Morphology

2009

BERTEIN Jean-Claude, CESCHI Roger
Discrete Stochastic Processes and Optimal Filtering / 2nd edition

CHANUSSOT Jocelyn *et al.*
Multivariate Image Processing

DHOME Michel
Visual Perception through Video Imagery

GOVAERT Gérard
Data Analysis

GRANGEAT Pierre
Tomography

MOHAMAD-DJAFARI Ali
Inverse Problems in Vision and 3D Tomography

SIARRY Patrick
Optimization in Signal and Image Processing

2008

ABRY Patrice *et al.*
Scaling, Fractals and Wavelets

GARELLO René
Two-dimensional Signal Analysis

HLAWATSCH Franz *et al.*
Time-Frequency Analysis

IDIER Jérôme
Bayesian Approach to Inverse Problems

MAÎTRE Henri
Processing of Synthetic Aperture Radar (SAR) Images

MAÎTRE Henri
Image Processing

NAIT-ALI Amine, CAVARO-MENARD Christine
Compression of Biomedical Images and Signals

NAJIM Mohamed
Modeling, Estimation and Optimal Filtration in Signal Processing

QUINQUIS André
Digital Signal Processing Using Matlab

2007

BLOCH Isabelle
Information Fusion in Signal and Image Processing

GLAVIEUX Alain
Channel Coding in Communication Networks

OPPENHEIM Georges *et al.*
Wavelets and their Applications

2006

CASTANIÉ Francis
Spectral Analysis

NAJIM Mohamed
Digital Filters Design for Signal and Image Processing